Invading Nature - Springer Series in Invasion Ecology

Volume 15

Series Editor
Daniel Simberloff, Department of Ecology and Evolutionary Biology
University of Tennessee, Knoxville, TN, USA

Biological Invasions represent one of those rare themes that cut across the disciplines of academic biology, while having profound environmental, philosophical, socioeconomic, and legislative implications at a global scale. There can be no doubt that biological invasions represent the single greatest threat to biodiversity past the activities of humankind itself. The implications are far reaching. Novel ecological and evolutionary forces are now directing the future expression of life itself, as native species and the communities that they comprise contend with invading species. The rules of the game have been suddenly and irrevocably changed.

Invading Nature - Springer Series in Invasion Ecology is a new book series topically spanning the breadth of invasion biology. The series is of singular importance as an integrative venue focusing on the broader ecological and evolutionary issues arising from non-native species, the impacts such species have in particular environs, trends patterns and processes, as well as causes and correctives. The series seeks novel and synthetic approaches to invasions including experimental, theoretical, systematic and conceptual treatments.

Prospective authors and/or editors should consult the **Series Editor Daniel Simberloff** for more details:

E-mail: tebo@utk.edu

Ruquia Gulzar • Anzar Ahmad Khuroo
Irfan Rashid

Field Manual on Alien Flora of Kashmir Himalaya

Casual, Naturalised and Invasive Plants

 Springer

Ruquia Gulzar 🆔
Department of Botany
University of Kashmir
Srinagar, India

Anzar Ahmad Khuroo 🆔
Department of Botany
University of Kashmir
Srinagar, India

Irfan Rashid 🆔
Department of Geoinformatics
University of Kashmir
Srinagar, India

ISSN 1874-7809   ISSN 2543-0483   (electronic)
Invading Nature - Springer Series in Invasion Ecology
ISBN 978-3-031-33849-6   ISBN 978-3-031-33847-2   (eBook)
https://doi.org/10.1007/978-3-031-33847-2

This Springer imprint is published by the registered company Springer Nature Switzerland AG
The registered company address is: Gewerbestrasse 11, 6330 Cham, Switzerland

Paper in this product is recyclable.

A panoramic view of plant invasion in the terrestrial ecosystem (Gulmarg)

A panoramic view of plant invasion in the aquatic ecosystem (Dal Lake)

Foreword

In the era of big data, large global consortia, open science, and sophisticated analytical tools, ecology has become fascinated by opportunities to search for general principles, ask questions that were hard to imagine a few decades ago and gain insights into processes that are applicable across taxa, scales and time. Research in biological invasions is no exception – the last decade has seen an outburst of papers exploring the mechanisms and processes determining the success of alien species at large scales, making use of dramatically improved information on alien species distributions, ecology, and traits. Such growing knowledge is paramount for better management and mitigation of the impacts of plant invaders. Still, it is good to be reminded that none of these achievements would have been possible without good old botanical field research. When it comes to real plants, you need to go to the field rather than just to databases and do some natural history.

The book **Field Manual on Alien Flora of Kashmir Himalaya: Casual, Naturalised and Invasive Plants** represents a brilliant example of such an approach where science meets practice. Authored by Ruquia Gulzar (a doctoral student), Anzar A. Khuroo (a botanist) and Irfan Rashid (a spatial ecologist) from the University of Kashmir, it provides profiles of the hundred alien plant species in Kashmir Himalaya with comprehensive information on their taxonomy, ecology, current invasion status, impacts, distribution map and photographs that are very helpful for field identification. Written to promote awareness among researchers, students, citizen scientists, stakeholders, and the general public on the threats invasive plants pose to the environment and biodiversity and encourage research and management, the book will be most beneficial for anyone interested in plant invasions in India and beyond.

Department of Invasion Ecology   Petr Pyšek
Institute of Botany, Czech Academy of Sciences
Průhonice, Czech Republic

Preface

This ***Field Manual on Alien Flora of Kashmir Himalaya: Casual, Naturalised and Invasive Plants*** has been prepared with two main goals in mind: to promote awareness on the threats of invasive alien species and to encourage research on and management of the common casual, naturalised, and invasive alien plants in Kashmir Himalaya, India. The *Manual* contains concise information on taxonomy, ecology, invasion status, impacts and distribution of 100 alien plant species of this Himalayan region. To facilitate field identification, the *Manual* provides photographs of habitat, habit, leaves, flowers and fruits of each plant species. The *Manual* is organised into three chapters:

Chapter 1 provides a brief account on the discipline of invasion biology right from its genesis to the present status. The basic concepts of the invasion process, mechanisms of invasion, its impacts, options for management and policy issues are discussed. The role and interlinkages of invasion biology in contemporary global environmental challenges of climate and land use changes and biodiversity loss are also highlighted.

Chapter 2 guides the readers on how to use the *Manual*. It first provides an overview of the study region, including its location, geology, climate and vegetation. It then describes the methods used to conduct the study based on which this *Manual* is prepared. It also describes the terminology and definitions used in the chapter 3 (Alien Species' Profile).

Chapter 3 comprises the bulk of this *Manual* and contains information on 100 alien plant species of the study region. For each species, the scientific name, name of its family, English name, local name, taxonomic characters, ecological traits, current invasion status (casual, naturalised and invasive) in Kashmir Himalaya, native range of the species, its global distribution, a regional distribution map and photo illustrations portraying diagnostic field characters are provided.

The *Manual* is the outcome of extensive field work conducted by the authors across this Himalayan region. It is hoped that the *Manual* will increase awareness on the threats posed by plant invasions among all the stakeholders—researchers, land managers, policy makers, environmentalists, naturalists, citizen scientists, students and the general public. It will also help in promoting research, formulating policies and planning management actions to deal with invasive alien species.

Srinagar, India   Ruquia Gulzar
Srinagar, India   Anzar Ahmad Khuroo
Srinagar, India   Irfan Rashid

Acknowledgements

Writing a book is a teamwork, because it involves inputs and help from many individuals and institutions. Same is true for this *Manual* as well.

We profusely acknowledge the valuable suggestions and insightful inputs received from Professor Daniel Simberloff, Series Editor, *Invading Nature – Springer Series in Invasion Ecology*. Right from initial conceptualisation of this book idea to its final publication, he has remained intimately associated with this work. We are beholden to Dr K. V. Sankaran, Consultant, FAO, for sparing time to go through the draft manuscript and suggest valuable comments and edits which has improved the flow in presentation. The expert suggestions by Dr Imtiyaz A. Lone, Govt Degree College Anantnag, on local names (Kashmiri) of the plants, and Dr Javaid Iqbal, South Campus, University of Kashmir, on proper usage of English names of plants are sincerely acknowledged. We are also grateful to the Gardeners working at Kashmir University Botanical Garden for their valuable inputs on local names of the plants. The generous help received from Mr. Akhtar Hussain Malik, other staff members (including Bilal A. Mir), and the research scholars (past and present members of BIOTA lab) at the Centre for Biodiversity and Taxonomy, Department of Botany, University of Kashmir, is highly acknowledged. Further, we record our gratitude to many individuals and institutions which we have not specifically named due to paucity of space, but whose inputs have directly or indirectly benefitted this work.

On personal behalf, I (Ruquia Gulzar) feel that no words will suffice to express the immense gratitude and admiration for my beloved parents (Mr. Gulam Mohammad and Mrs. Sarwa) for their unconditional love, unwavering support and unflinching faith during the course of this work. I am highly grateful to my siblings (Mir Shahnawaz, Roohi) and niece (Sehrish) for their support and kind help during the field surveys. I am greatly indebted to my supervisor (Anzar Ahmad Khuroo) for his proficient guidance, unparalleled encouragement and timely help. I am also thankful to my co-supervisor (Irfan Rashid) for his valuable suggestions and timely help. I would also like to acknowledge the financial support received from CSIR-New Delhi during the present study.

Likewise, Anzar Ahmad Khuroo personally appreciates his better half (Mohsina), kids (Adeena, Sufwan, Mansha) and parents for allowing him to quite often spare home and holiday time on this work, which was otherwise genuinely due to them. Looking back, Anzar would put on record his deep sense of indebtedness to Prof G. H. Dar and Prof. Z. A. Reshi, University of Kashmir, for their scientifically stimulating mentorship: their taxonomic and ecological 'footprints' are reflected in this work. Anzar would also like to duly acknowledge the support received over the last decade from various grant agencies (SAC-ISRO, Ahmedabad, MoEF&CC, SERB-DST and DBT, New Delhi, India), which has helped this book in one way or the other.

We are highly obliged to *Springer Nature* and their professional team, especially Eva Loerinczi, Anette Weiss, Rajeshree, Kayalvizhi for coordinating the manuscript processing, proofing and printing. Last but not least, no book can ever claim to be complete and free from errors (factual or typographical), and the same applies to this work as well. Therefore, it is our humble request to the esteemed readers of this book to share their valuable feedback with the authors.

Contents

1 Invasion Biology 1
   1.1 Introduction 1
   1.2 Biological Invasions: A Multi-stage Process 2
   1.3 Species Invasiveness and Habitat Invasibility 3
   1.4 Impacts of Biological Invasions 4
   1.5 Management and Policymaking 6
   1.6 Relevance of the Manual 8
   References 9

2 How to Use the Field Manual? 13
   2.1 Study Region 13
   2.2 Methods Used 14
   2.3 Terminology and Definitions 15
   References 20

3 Alien Species' Profile 23

Index 225

About the Authors

Ruquia Gulzar doctoral student in the Department of Botany at the University of Kashmir, is working on the 'Naturalised Flora of Kashmir Himalaya' for the last five years. In pursuit of her research, she has conducted extensive field surveys in this Himalayan region and laboratory-based studies. Recently, she has also published her research work in reputed scientific journals (see at https://www.researchgate.net/profile/Ruquia-Gulzar).

Anzar Ahmad Khuroo Associate Professor in the Department of Botany at the University of Kashmir, is a botanist with two decades-long research experience on plant biodiversity in Kashmir Himalaya. One of his focus areas of research is inventory, impacts and management of invasive alien flora. He has made pioneering research contribution on alien flora of the Himalaya in particular and India in general. He has more than 180 papers and a book *Biodiversity of the Himalaya: Jammu and Kashmir State* to his credit (see https://botany.uok.edu.in/Main/ProfilePage.aspx?Profile=098).

Irfan Rashid Senior Assistant Professor in the Department of Geoinformatics at the University of Kashmir, is a spatial ecologist with a decade-long research experience on natural resource mapping in the Himalaya. He has extensively published on the Himalayan environment (see https://geoinformatics.uok.edu.in/Main/ProfilePage.aspx?Profile=0207).

Chapter 1
Invasion Biology

1.1 Introduction

We are currently living in an age of Anthropocene [1]. Amongst the dominant drivers of Anthropocene, biological invasions by alien species are recognised as the second greatest threat to global biodiversity after habitat loss [2]. In recent times, the rising global trade, travel and transport have accelerated the rate of species' introductions outside their native biogeographic ranges [3]. Worldover, a significant proportion of the introduced species pool has become naturalised and some of these spread as invasive in their non-native ranges, causing significantly high ecological and economic impacts affecting nature and nature's contribution to all life forms [4, 5]. In response, concerted efforts are being made across the globe to focus scientific research, strengthen biosecurity, formulate policies and implement management strategies to combat biological invasions [6–8].

Invasion biology—a scientific discipline that deals with the study of invasions by alien species—has its origin in the publication of Charles Elton's seminal book *The Ecology of Invasions by Animals and Plants* [9]. Since then, the discipline has emerged with global research efforts focussed on understanding the patterns and processes of biological invasions, impacts of invasive alien species on biodiversity and ecosystem functioning, and management and restoration strategies [10]. More specifically, a rapid expansion of the discipline over the last two decades reflects its increasing relevance in addressing the global crises of biodiversity loss, climate change and very recently the COVID-19 pandemic [2, 3, 11]. Invasion biology, as an interdisciplinary subject, has embraced a wide range of domains in ecology and related areas (e.g. biogeography, population biology, evolutionary biology, physiology and genomics) and also found interlinkages with divergent disciplines such as resource economics, epidemiology, risk analysis, vector science and biosecurity [12, 13]. More recently, the ongoing climate and land-use changes have contributed

R. Gulzar et al., *Field Manual on Alien Flora of Kashmir Himalaya*,
Invading Nature - Springer Series in Invasion Ecology 15,
https://doi.org/10.1007/978-3-031-33847-2_1

to the growing complexity of biological invasions and their ecological and economic fallouts [14]. Along with these new challenges, an ever-expanding connection with diverse disciplines has fostered the transdisciplinary growth of invasion biology [2, 15].

1.2   Biological Invasions: A Multi-stage Process

As a process, biological invasion involves a series of stages that all alien species may have to pass through along the invasion continuum. The main stages include import or introduction of alien species into a non-native range, its release or escape into the wild, propagation of its population without human intervention followed by its spread as invasive. These multiple stages can be characterised along a continuum known as introduction–naturalisation–invasion continuum [16, 17]. Recognising the status of an alien species at these stages is critical to comprehend how a variety of factors facilitate species transition over the continuum and what management action is suitable at each stage. Therefore, these stages of invasion process are reflected whilst characterising status of each species in alien inventories [18]. On entering into a non-native region, an alien species has to survive, reproduce, disperse and spread to be finally recognised as an invasive [16]. For instance, in case of plants, those alien species which grow outside the area of their cultivation but need human assistance directly or indirectly to survive are known as casual alien species. And, those alien species which form self-sustaining populations in the wild without human assistance are known as naturalised, and those species that spread widely across a region with impacts are known as invasive [19, 20].

Generally, there are three main mechanisms of introduction: importation of goods, influx of a transport vector and natural spread from an adjoining region where the species is an alien [15]. International trade and travel are considered the principal drivers of biological invasions in both terrestrial and aquatic ecosystems [21, 22]. The traded commodity can be an alien species (e.g. ornamental plants) in some cases but, most often, alien species (or their propagules) are introduced unintentionally as an impurity or contaminant of traded commodities, e.g. weed seeds in grain, parasites in livestock or as a runaway on transport vessels or other means of transportation [23], and ballast water and hull fouling are the main pathways/vectors for marine invasions.

After successful introduction and escape into the wild (human-dominated and/or natural ecosystems) in non-native regions, a proportion of the alien species successfully establishes and enters into the naturalised stage. The time to naturalisation can vary from years to decades depending upon the species' lag phase and pace of propagule pressure. Although majority of the imported species fail to establish long-term populations, a considerable proportion do so, and the number of such species is steadily growing globally [3]. Naturalisation is the most crucial stage in the invasion process because all naturalised species are potential future invaders. The naturalised species pool is often recognised as the most important proxy for estimating invasion debt in an area [24]. The number of naturalised species varies greatly throughout the world, and insightful biogeographical patterns have emerged [25]. Recent studies show that introduced species on islands are more frequently naturalised than those on the terrestrial lands [26]. The number of naturalised species on islands shows an increase with increasing temperature since it aids propagule establishment [27]. Similarly, the number of naturalised species in temperate zones is reported to decline with latitude whilst their geographical ranges increase. The final step for alien species is the transition from naturalised to invasive stage.

The stage-based approach in biological invasions has a fundamental and practical relevance in better understanding how alien species spread in non-native ranges, why some ecosystems are more susceptible to invasion, what factors determine invasive species impacts, and which species and sites/habitats are to be prioritised for management action [28]. Identifying each stage of the invasion process and recognising the status of alien species pool is thus crucial because each stage is mediated by a separate set of causes and factors, some of which may operate at multiple stages [27]. The importance of factors at each stage can vary, with socio-economic factors being the most important at earlier stages, followed by biogeographical, ecological, and evolutionary factors. But, all these factors can influence all stages of invasion along the invasion continuum [29]. Some of the factors can be ecological such as availability of a breeding partner, the need for specialised dispersers, obligatory mutualism, as well as plant traits (e.g. lifeform); likewise, the human-mediated factors include mainly introduction of the species.

1.3 Species Invasiveness and Habitat Invasibility

Species invasiveness and habitat invasibility are two fundamental conceptual pillars of invasion biology [30]. Species invasiveness is defined as the ability of an alien species to invade the recipient ecosystem and the major determinants include introduction history, species traits, evolutionary, and ecological factors [31]. On the other hand, vulnerability of the recipient ecosystem to the establishment and spread of an alien species is referred to as habitat invasibility. A number of hypotheses have been posited to investigate these twin concepts of invasion biology. One of the most extensively investigated mechanisms of alien plant invasions is the enemy release hypothesis (ERH), which hypothesises that, when plant species are introduced to a non-native environment, their control by herbivores and other natural predators is reduced, leading to a fast expansion in range and abundance [32, 33]. This hypothesis is built on a three-point argument: (1) natural enemies are significant regulators of plant populations; (2) enemies have a higher impact on native species than invasive species; and (3) plants can benefit from reduced enemy regulation, thereby leading to greater population expansion. Depending on the species under investigation, the reliability of these arguments and the likelihood for enemy release also varies [33].

In addition, there are several other factors that enable alien plants to spread quickly in introduced ranges which include: capability to reproduce both sexually and asexually, popularity as ornamentals, resistance to harsh environmental conditions, high regeneration, early maturation and effective seed dispersal modes. Depending upon the magnitude of human disturbance regimes and climate change, the adaptability and/or spreading potential of the alien plants may vary [34]. The evolutionary past of a species and its environment determine its ability to adapt to changing climates and dispersal in the non-native region. For instance, the floral diversity of the Western Himalaya has been shown to have a slower rate of colonisation and altitude-facilitated speciation, resulting in a predominantly native species composition in the high-altitude Himalaya [35]. The native plants are unable to survive in the altered environment or long-distance propagation to the new habitat as a result of this slower colonisation potential and climate-specific speciation. On the other hand, invasive alien plants with continental distributions may benefit from global environmental changes due to their physiological adaptability and possibility of acquiring new dispersal agents [36]. Also, land-use changes are responsible for altering native biodiversity patterns as novel ecosystems with combinations of alien and native species are becoming common throughout the world [37].

1.4 Impacts of Biological Invasions

Invasive alien species have the ability to inflict significant negative consequences on the environment, economy, and human health. These species are key drivers of change in recipient ecosystems, causing changes in ecosystem functioning and biodiversity [38]. Disturbed habitats like grazing pastures, open meadows, roadsides, and woodland openings are reported to be highly susceptible to invasive species worldwide [39, 40]. Also, these species pose a significant risk to the endemic biodiversity and life-supporting ecosystem services in the invaded biomes [41, 42]. In extreme cases, invasive alien species may cause extinction of native species through predation or competition for habitat and food [43]. Invasive species lead to a decline in species richness in both terrestrial and aquatic ecosystems. A global meta-analysis evaluated the role of invasive alien species in reducing native species richness and found that even a single invasive species can result in 16.6% reduction in native species richness [44]. The most significant effect on species richness and evenness is caused by tall invasive species that can establish populations with a cover noticeably greater than that of native dominating species [45]. Besides impacts on aboveground part of terrestrial ecosystems, plant invasions affect belowground soil system dynamics as well. Understanding the invasion impacts on belowground part of ecosystems is crucial for the effective management and restoration of invaded landscapes. The documented impacts of invasive alien species on soil systems are varied and include changes in cycling of carbon, nitrogen, and other nutrient pools, as well as modification of physio-chemical properties like soil pH, moisture, temperature, organic matter, and microbial activity [41, 42, 46]. Invasive alien species can also cause the transmission and spread of diseases in agriculture and forestry sectors [47, 48]. Therefore, management expenditures for the eradication of invasive alien species to mitigate damage can have considerable financial implications. The Generic Impact Scoring System (GISS) and the Environmental Impact Classification for Alien Taxa (EICAT) are uniform standardised approaches for evaluating, comparing, and eventually predicting the magnitude of various consequences of invasive alien species at a global scale [49].

Research attention has recently been focused on the economic costs of biological invasions, resulting in a rapid accumulation of evidence on invasion-related economic losses [50]. Based on a novel database viz., InvaCost, economic costs of biological invasions across the globe were recently estimated [4]. The study revealed that the global economic costs of invasive alien species amounted to at least US$ 1.3 trillion during the period 1970–2017, and the costs are increasing rapidly. The urgency of preventing and controlling biological invasions is highlighted by such huge costs, which are probably an underestimate, particularly for invasive alien plants [50]. The study by Diagne et al. [4] significantly underestimated the costs due to plant invasions compared to vertebrate and invertebrate invasions. It was estimated that, globally, between 1970 and 2017, invasive alien plants cost US$ 8.9 billion out of a total of US$ 591 due to plants, vertebrates, and invertebrates. However, the cost was more than US$ 8.9 billion for plants based on information available over the past 20 years [50]. For example, invasive alien plants are estimated to cost at least € 3.8 billion a year in Europe, accounting for 30% of the continent's total costs due to invasive alien species [51].

In an era of global environmental change, ecological novelty as an element of species redistribution is considered one of the effects of biological invasions [52]. Different dimensions of ecological novelty have been characterised and linked to range-expanding alien species in multiple ways. Ecological novelty can be quantified inversely as the eco-evolutionary process of interaction between native and alien species. [53]. The alien species with new weapons or other novel qualities that the native species is unfamiliar with (i.e. limited eco-evolutionary experience) are likely to have greater impacts on native species than those alien species that are comparable to native ones in these traits. The alien mammalian predators that were introduced into New Zealand (and other oceanic islands) had a highly deleterious impact on local flightless birds, with Kiwi and other soil-nesting bird species becoming easy targets for mustelids and cats since these predators were absent in the region previously [54]. Similarly, range-expanding plants that are spreading into biomes with colder climates in response to climate change can have significant effects on nutrient cycling and ecosystem functioning when alien plants with new growth forms unfamiliar to the region are introduced. For example, laurophyllisation occurs in temperate forests and shrubification in tundra ecosystems [53].

Whilst most of the research has focussed on the detrimental impacts of biological invasions [55], the possible advantages, though less known, pose formidable management challenges [56]. Therefore, assessing the impacts of alien species is crucial from a scientific standpoint for successful implementation of management strategies. To make objective management decisions, one must adopt a perspective that takes into account both ecosystem services and disservices due to invasive alien species.

1.5 Management and Policymaking

Globally, recent research has focussed more towards management of invasive alien species and ecological restoration of invaded landscapes. As already emphasised, targeting the stage of invasion is critical to the designing and application of successful management methods [57]. Invasive alien species management plans must include strategies related to prevention, early detection and rapid response (EDRR), containment and control at specific stages along the invasion continuum in order to be effective and efficient [58]. As majority of invasive plants are introduced intentionally, the most effective strategy to avoid future introductions is to prevent them from getting introduced in the first place. Robust tools of risk analysis (RA), which assess the biology of the species, the characteristics of the ecosystem into which it is being introduced, and whether it has been reported as invasive somewhere, help to select suitable species for introduction and prevent unsuitable species. If risk analysis fails to prevent the arrival of an invasive or potentially invasive species, it is vital that it be recognised early at the point of entry and removed before it gains entry. In EDRR, the term early detection is defined as the act of surveillance for already introduced alien species, and rapid response as any action that allows those organisms to be destroyed or stopped them from spreading further [59]. For a rapid response to be successful, eradication operations must be completed within weeks or at most 1–2 years, because each invasion scenario is different and the timescale for eradication varies [60]. Therefore, developing and implementing a proactive surveillance plan is critical. If monitoring does not result in the timely identification of a potentially harmful alien plant and removal is no longer an option, containment of the species is another option. If this also fails due to its vast and abundant distribution of the species, a control method must be implemented. Cultural, physical, chemical or biological control or a mix of some or all of these, could be used in a control plan, which should then be followed by restoration [61]. Of late, species distribution models (SDM) have emerged as commonly used tools to evaluate invasion risk and, in most cases, to predict species habitat suitability [41, 42, 62, 63]. However, applicability of SDMs to decipher patterns of abundance and impact may be limited [64]. Recently, citizen science has shown a huge promise in early detection of potential invasive species and map their occurrence thus helping in planning invasion management [65]. The use of smartphone-based apps for reporting and uploading photos and geospatial information of potential invasive alien species at their early stage of introduction has improved chances of early detection and rapid control of new invasive alien species [66].

Worldover, with nations having almost failed to meet the Aichi Biodiversity Target 2020, negotiations are being held under the aegis of the Convention on Biological Diversity (CBD) to develop a feasible and monitorable Post-2020 Global Biodiversity Framework. As per this Framework, to meet the 2050 Vision, where biodiversity is respected, maintained, and restored, we must reduce the significant and still rising impact of invasive alien species (IAS) on biodiversity. To achieve this, the IUCN Invasive Species Specialist Group (ISSG) has established a preliminary target that was formally presented to the CBD for discussion on Post-2020 targets [67]. The Aichi Target 9 of CBD related to IAS aimed to 'control all pathways for the introduction of invasive alien species, achieving by 2030 a 50% reduction in the rate of new introductions, and eradicate or control invasive alien species to eliminate or reduce their impacts by 2030 in at least 50% of priority sites' [68]. Reducing the effects caused by IAS has been identified as a priority area of action by various global environmental policies. The United Nations Sustainable Development Goals (SDGs) also include an IAS target that intends to 'avoid the introduction and considerably minimise the effect of IAS in terrestrial and aquatic ecosystems, as well as control or eradicate priority species'. The number of countries passing appropriate legislation and formally establishing institutional mechanisms for invasion management is used to gauge the progress.

Globally, significant progress has been made in gathering data and projecting alien species distributions [31, 69]. As a result, various alien species databases, ranging in scale from regional to continental, have been developed in the recent decade, with the goal of gathering data on alien species' status, distribution, pathways, impacts, and other characteristics [70]. For example, the European DAISIE project was ground-breaking in this respect since it included all important species and biomes for an entire continent and the data was collected using a uniform criterion [69]. Open access to data on the distribution of alien species, on the other hand, remains a challenge, particularly for countries that need to use such information for developing policies, plan management and surveillance. Also, the scientific literature on biological invasions is mostly available from developed nations, with relatively little known from developing nations. Global syntheses and inferences on the diversity and distribution of IAS and their impacts are often drawn based on the fragmentary datasets, though the huge knowledge gaps from developing countries are well known [71]. In fact, effective management measures for IAS, may be more necessary and advantageous for developing nations, where larger, and highly diverse natural ecosystems with greater biodiversity are more common [72]. Lack of skills and technical knowhow, little awareness about the scale of problem, inadequate policies and imperfect implementation of regulations are the main reasons why developing economies are not successful in combating IAS threats. Therefore, focussing on research and generation of appropriate knowledge products to inform policy and manage biological invasions in developing countries, particularly emerging economies like India, assumes immediate priority.

1.6   Relevance of the Manual

Mountains, with 12.5% of the Earth's land surface and covering 13.8 million km^2, are home to one-quarter of all terrestrial species [73]. Mountains are a natural treasure trove of precious ecosystem goods and services and progenitors of biodiversity and cradles of evolution [74]. Because of recent climatic and land-use changes, urban population growth, road construction, expansion of agricultural land and a host of unsustainable human activities, the natural habitats in mountains, across the globe, that were previously presumed to be resistant to biological invasion are now increasingly becoming susceptible to invasion by alien species [73, 75–77]. Amongst the mountain systems of the world, the Himalaya, endowed with a rich repository of biological diversity and recognised as a global biodiversity hotspot, occupies a prominent place [78]. The Himalaya is home to hydrological resources supporting nearly two billion human population in South Asia. Although spread across different countries in South Asia, the major portion of the Himalaya falls within India. The Kashmir Himalaya, located towards northwestern side of Indian Himalaya is also one of the most eco-fragile mountainous landscapes of the world [79]. Studies indicate that changing land-use patterns and rapidly warming climate have increased the risk of biological invasions in this Himalayan region also [80–82]. Although several studies on different aspects of plant invasions have been carried out in this region [39, 41, 42, 83], there is a dearth of information that could help rapid and reliable field identification and characterisation of alien plant species in the region. Against this background, publication of an illustrated *Manual* with a focus on field-based taxonomic characters, species traits and regional distribution is timely. It is expected to play a pivotal role in creating awareness on biological invasions, advancing research and planning management actions.

In fact, Ricciardi et al. [84] have recently highlighted that our capacity to assess and comprehend the spatiotemporal dynamics of invasions and their effects is critically hampered by the taxonomic barrier. A large portion of the current knowledge is based on secondary data syntheses and analyses using information from regional checklists and atlases of flora. In all such cases, the reliability of species identifications has substantially impeded their utility [85]. Misidentifications and inability to recognise cryptic species complexes have frequently prevented the development and use of effective control measures [86]. The present *Manual* is expected to function as a reliable source of information for developing invasion management, policy and practice [87]. Such manuals or field guides are available for some regions of the world, particularly in the developed world [88, 89]. However, to the best of our knowledge, there is no such field manual available for the Himalaya in particular and very few for the developing world in general.

References

1. Crutzen PJ. The "anthropocene". In: Earth system science in the anthropocene. Berlin: Springer; 2006. p. 13–8.
2. Pyšek P, Hulme PE, Simberloff D, Bacher S, Blackburn TM, Carlton JT, Richardson DM. Scientists' warning on invasive alien species. Biol Rev. 2020;95(6):1511–34.
3. Seebens H, Blackburn TM, et al. No saturation in the accumulation of alien species worldwide. Nat Commun. 2017;8:14435. https://doi.org/10.1038/ncomms14435.
4. Diagne C, Leroy B, Vaissière AC, Gozlan RE, Roiz D, Jarić I, Salles JM, Bradshaw CJ, Courchamp F. High and rising economic costs of biological invasions worldwide. Nature. 2021a;592(7855):571–6.
5. Simberloff D, Martin JL, Genovesi P, Maris V, Wardle DA, Aronson J, Courchamp F, Galil B, García-Berthou E, Pascal M, Pyšek P. Impacts of biological invasions: what's what and the way forward. Trends Ecol Evol. 2013;28(1):58–66.
6. Diagne C, Leroy B, Gozlan RE, Vaissière AC, Assailly C, Nuninger L, et al. InvaCost, a public database of the economic costs of biological invasions worldwide. Scientific data. 2020;7(1):1–12.
7. Foxcroft LC, Wilgen BWV, Abrahams B, Esler KJ, Wannenburgh A. Knowing-doing continuum or knowing-doing gap? Information flow between researchers and managers of biological invasions in South Africa. In: Biological invasions in South Africa. Cham: Springer; 2020. p. 831–53.
8. Morelli TL, Brown-Lima CJ, Allen JM, Beaury EM, Fusco EJ, Barker-Plotkin A, et al. Translational invasion ecology: bridging research and practice to address one of the greatest threats to biodiversity. Biol Invasions. 2021;23(11):3323–35.
9. Elton CS. The ecology of invasions by animals and plants. 2nd ed. Chicago: Springer; 1958. p. 181. https://doi.org/10.1007/978-94-009-5851-7.
10. Ricciardi A, MacIsaac HJ. The book that began invasion ecology. Nature. 2008;452(7183):34.
11. Zhang L, Rohr J, Cui R, Xin Y, Han L, Yang X, et al. Biological invasions facilitate zoonotic disease emergences. Nat Commun. 2022;13(1):1–11.
12. Ricciardi A, Cassey P, Leuko S, Woolnough AP. Planetary biosecurity: applying invasion science to prevent biological contamination from space travel. Bioscience. 2022;72(3):247–53.
13. Vaz AS, Kueffer C, Kull CA, Richardson DM, Schindler S, Muñoz-Pajares AJ, et al. The progress of interdisciplinarity in invasion science. Ambio. 2017;46(4):428–42.
14. Robinson, T. B., Martin, N., Loureiro, T. G., Matikinca, P., & Robertson, M. P. Double trouble: the implications of climate change for biological invasions. NeoBiota. 2020;62:463 487.
15. Richardson DM, Pyšek P, Carlton JT. A compendium of essential concepts and terminology in invasion ecology. In: Fifty years of invasion ecology: the legacy of Charles Elton, vol. 1; 2011. p. 409–20.
16. Blackburn TM, Pyšek P, Bacher S, Carlton JT, Duncan RP, Jarošík V, et al. A proposed unified framework for biological invasions. Trends Ecol Evol. 2011;26(7):333–9.
17. Richardson DM, Pyšek P, Rejmánek M, Barbour MG, Panetta FD, West CJ. Naturalization and invasion of alien plants: concepts and definitions. Divers Distrib. 2000;6(2):93–107.
18. Khuroo AA, Reshi ZA, Malik AH, Weber E, Rashid I, Dar GH. Alien flora of India: taxonomic composition, invasion status and biogeographic affiliations. Biol Invasions. 2012;14(1):99–113.
19. Pyšek P, Pergl J, Essl F, Lenzner B, Dawson W, Kreft H, et al. Naturalized alien flora of the world: species diversity, taxonomic and phylogenetic patterns, geographic distribution and global hotspots of plant invasion. Preslia. 2017;89:203–74.
20. Pyšek P, Richardson DM, Rejmánek M, Webster GL, Williamson M, Kirschner J. Alien plants in checklists and floras: towards better communication between taxonomists and ecologists. Taxon. 2004;53(1):131–43.
21. Dawson W, Burslem DF, Hulme PE. Factors explaining alien plant invasion success in a tropical ecosystem differ at each stage of invasion. J Ecol. 2009;97(4):657–65.

22. Essl F, Bacher S, Blackburn TM, Booy O, Brundu G, Brunel S, et al. Crossing frontiers in tackling pathways of biological invasions. Bioscience. 2015;65(8):769–82.
23. Hulme PE. Unwelcome exchange: international trade as a direct and indirect driver of biological invasions worldwide. One Earth. 2021;4(5):666–79.
24. Essl F, Dullinger S, Rabitsch W, Hulme PE, Hülber K, Jarošík V, et al. Socioeconomic legacy yields an invasion debt. Proc Natl Acad Sci. 2011;108(1):203–7.
25. Pyšek P, Richardson DM, Jarošík V. Who cites who in the invasion zoo: insights from an analysis of the most highly cited papers in invasion ecology. Preslia. 2006;78:437–68.
26. Stroud JT. Island species experience higher niche expansion and lower niche conservatism during invasion. Proc Natl Acad Sci. 2021;118(1)
27. Catford JA, Vesk PA, Richardson DM, Pyšek P. Quantifying levels of biological invasion: towards the objective classification of invaded and invasible ecosystems. Glob Chang Biol. 2012;18(1):44–62.
28. Theoharides KA, Dukes JS. Plant invasion across space and time: factors affecting nonindigenous species success during four stages of invasion. New Phytol. 2007;176(2):256–73.
29. Williamson M. Explaining and predicting the success of invading species at different stages of invasion. Biol Invasions. 2006;8(7):1561–8.
30. Richardson DM, Pyšek P. Plant invasions: merging the concepts of species invasiveness and community invasibility. Prog Phys Geogr. 2006;30(3):409–31.
31. van Kleunen M, Pyšek P, Dawson W, Kreft H, Pergl J, Weigelt P, et al. The global naturalized alien flora (GloNAF) database. Ecology. 2019;2019(100):1.
32. Heger T, Jeschke JM. The enemy release hypothesis as a hierarchy of hypotheses. Oikos. 2014;123(6):741–50.
33. Keane RM, Crawley MJ. Exotic plant invasions and the enemy release hypothesis. Trends Ecol Evol. 2002;17(4):164–70.
34. Hampe A, Petit RJ. Conserving biodiversity under climate change: the rear edge matters. Ecol Lett. 2005;8(5):461–7.
35. Mani MS. Biogeographical evolution in India. In: Ecology and biogeography in India. Dordrecht: Springer; 1974. p. 698–724.
36. Mungi NA, Coops NC, Ramesh K, Rawat GS. How global climate change and regional disturbance can expand the invasion risk? Case study of Lantana camara invasion in the Himalaya. Biol Invasions. 2018;20(7):1849–63.
37. Mazía N, Chaneton EJ, Ghersa CM. Disturbance types, herbaceous composition, and rainfall season determine exotic tree invasion in novel grassland. Biol Invasions. 2019;21(4):1351–63.
38. Jochum M, Ferlian O, Thakur MP, Ciobanu M, Klarner B, Salamon JA, et al. Earthworm invasion causes declines across soil fauna size classes and biodiversity facets in northern north American forests. Oikos. 2021a;130(5):766–80.
39. Khuroo AA, Malik AH, Reshi ZA, Dar GH. From ornamental to detrimental: plant invasion of *Leucanthemum vulgare* Lam (Ox-eye Daisy) in Kashmir valley, India. Curr Sci. 2010;98(5):600–2.
40. Stutz S, Mráz P, Hinz HL, Müller-Schärer H, Schaffner U. Biological invasion of oxeye daisy (*Leucanthemum vulgare*) in North America: pre-adaptation, post-introduction evolution, or both? PLoS One. 2018;13(1):e0190705.
41. Ahmad R, Khuroo AA, Hamid M, Rashid I. Plant invasion alters the physico-chemical dynamics of soil system: insights from invasive *Leucanthemum vulgare* in the Indian Himalaya. Environ Monit Assess. 2019a;191(3):1–15.
42. Ahmad R, Khuroo AA, Hamid M, Charles B, Rashid I. Predicting invasion potential and niche dynamics of *Parthenium hysterophorus* (congress grass) in India under projected climate change. Biodivers Conserv. 2019b;28(8):2319–44.
43. Andriantsoa R, Jones JP, Achimescu V, Randrianarison H, Raselimanana M, Andriatsitohaina M, et al. Perceived socio-economic impacts of the marbled crayfish invasion in Madagascar. PLoS One. 2020;15(4):e0231773.
44. Mollot G, Pantel JH, Romanuk TN. The effects of invasive species on the decline in species richness: a global meta-analysis. In: Advances in ecological research, vol. 56. Academic Press; 2017. p. 61–83.

45. Hejda M, Pyšek P, Jarošík V. Impact of invasive plants on the species richness, diversity and composition of invaded communities. J Ecol. 2009;97(3):393–403.
46. Cuda J, Vítková M, Albrechtová M, Guo WY, Barney JN, Pyšek P. Invasive herb Impatiens glandulifera has minimal impact on multiple components of temperate forest ecosystem function. Biol Invasions. 2017;19(10):3051–66.
47. Liebhold AM, Brockerhoff EG, Kalisz S, Nuñez MA, Wardle DA, Wingfield MJ. Biological invasions in forest ecosystems. Biol Invasions. 2017;19(11):3437–58.
48. Young HS, Parker IM, Gilbert GS, Guerra AS, Nunn CL. Introduced species, disease ecology, and biodiversity–disease relationships. Trends Ecol Evol. 2017;32(1):41–54.
49. Sohrabi S, Pergl J, Pyšek P, Foxcroft LC, Gherekhloo J. Quantifying the potential impact of alien plants of Iran using the generic impact scoring system (GISS) and environmental impact classification for alien taxa (EICAT). Biol Invasions. 2021;23:2435–49. https://doi.org/10.1007/s10530-021-02515-6.
50. Novoa A, Moodley D, Catford JA, Golivets M, Bufford J, Essl F, et al. Global costs of plant invasions must not be underestimated. NeoBiota. 2021;69:75–8.
51. Kettunen M, Genovesi P, Gollasch S, Pagad S, Starfinger U, ten Brink P, Shine C. Technical support to EU strategy on invasive alien species (IAS) – assessment of the impacts of IAS in Europe and the EU. London: Institute for European Environmental Policy; 2009.
52. Saul WC, Jeschke J, Heger T. The role of eco-evolutionary experience in invasion success. NeoBiota. 2013;17:57.
53. Essl F, Dullinger S, Genovesi P, Hulme PE, Jeschke JM, Katsanevakis S, et al. A conceptual framework for range-expanding species that track human-induced environmental change. Bioscience. 2019;69(11):908–19.
54. Blackwell GL. Another world: the composition and consequences of the introduced mammal fauna of New Zealand. Australian Zoologist. 2005;33(1):108–18.
55. Bellard C, Genovesi P, Jeschke JM. Global patterns in threats to vertebrates by biological invasions. Proc R Soc B Biol Sci. 2016;283(1823):20152454.
56. Sheergojri IA, Rashid I, Rehman IU. Invasive species services-disservices conundrum: a case study from Kashmir Himalaya. J Environ Manag. 2022;309:114674.
57. Khuroo AA, Reshi Z, Rashid I, Dar GH, Khan ZS. Operational characterization of alien invasive flora and its management implications. Biodivers Conserv. 2008;17(13):3181–94.
58. Tobin PC. Managing invasive species. F1000Research. 2018:7.
59. Reaser JK, Burgiel SW, Kirkey J, Brantley KA, Veatch SD, Burgos-Rodríguez J. The early detection of and rapid response (EDRR) to invasive species: a conceptual framework and federal capacities assessment. Biol Invasions. 2020;22(1):1–19.
60. Lodge DM, Williams S, MacIsaac HJ, et al. Biological invasions: recommendations for US policy and management. Ecol Appl. 2006;16(6):2035–54.
61. Dusz MA, Martin FM, Dommanget F, Petit A, Dechaume-Moncharmont C, Evette A. Review of existing knowledge and practices of tarping for the control of invasive knotweeds. Plan Theory. 2021;10(10):2152.
62. Ncube B, Shekede MD, Gwitira I, Dube T. Spatial modelling the effects of climate change on the distribution of Lantana camara in southern Zimbabwe. Appl Geogr. 2020;117:102172.
63. Tiwari S, Mishra SN, Kumar D, Kumar B, Vaidya SN, Ghosh BG, et al. Modelling the potential risk zone of *Lantana camara* invasion and response to climate change in eastern India. Ecol Process. 2022;11(1):1–13.
64. Jarnevich CS, Sofaer HR, Engelstad P. Modelling presence versus abundance for invasive species risk assessment. Divers Distrib. 2021;27(12):2454–64.
65. Crall AW, Jarnevich CS, Young NE, Panke BJ, Renz M, Stohlgren TJ. Citizen science contributes to our knowledge of invasive plant species distributions. Biol Invasions. 2015;17(8):2415–27.
66. Kobori H, Dickinson JL, Washitani I, Sakurai R, Amano T, Komatsu N, et al. Citizen science: a new approach to advance ecology, education, and conservation. Ecol Res. 2016;31(1):1–19.
67. Essl F, Latombe G, Lenzner B, Pagad S, Seebens H, Smith K, et al. The convention on biological diversity (CBD)'s Post-2020 target on invasive alien species–what should it include and how should it be monitored? NeoBiota. 2020;62:99.

68. CBD 2020 available on https://www.cbd.int/gbo5. Accessed on 25-10-2021.
69. Hulme PE, editor. Handbook of alien species in Europe (Vol. 569). Dordrecht: Springer; 2009.
70. Hulme PE, Weser C. Mixed messages from multiple information sources on invasive species: a case of too much of a good thing? Divers Distrib. 2011;17(6):1152–60.
71. Pyšek P, Richardson DM, Pergl J, Jarošík V, Sixtová Z, Weber E. Geographical and taxonomic biases in invasion ecology. Trends Ecol Evol. 2008;23(5):237–44.
72. Nunez MA, Pauchard A. Biological invasions in developing and developed countries: does one model fit all? Biol Invasions. 2010;12(4):707–14.
73. Manish K. Species richness, phylogenetic diversity and phylogenetic structure patterns of exotic and native plants along an elevational gradient in the Himalaya. Ecol Process. 2021;10(1):1–13.
74. Hoorn C, Perrigo A, Antonelli A, editors. Mountains, climate and biodiversity. John Wiley & Sons; 2018.
75. Pauchard A, Kueffer C. Dietz H et al Ain't no mountain high enough: plant invasions reaching new elevations. Front Ecol Environ. 2009;7(9):479–86.
76. Averett JP, McCune B, Parks CG, Naylor BJ, et al. Non-native plant invasion along elevation and canopy closure gradients in a middle Rocky Mountain ecosystem. PloS one. 2016;11(1):e0147826.
77. McDougall KL, Alexander JM, Haider S, Pauchard A, Walsh NG, Kueffer C. Alien flora of mountains: global comparisons for the development of local preventive measures against plant invasions. Divers Distrib. 2011;17(1):103–11.
78. Basnet D, Kandel P, Chettri N, Yang Y, Lodhi MS, Htun NZ, et al. Biodiversity research trends and gaps from the confluence of three global biodiversity hotspots in the far-eastern Himalaya. Int J Ecol. 2019;2019:1–14.
79. Dar GH, Khuroo AA. *Biodiversity of the Himalaya: Jammu and Kashmir state* 263, vol. 18. Singapore: Springer Nature; 2020. p. 1–13.
80. Khuroo AA, Ahmad R, Hamid M, Rather ZA, Malik AH, Rashid I. An annotated inventory of invasive alien Flora of India. Invasive Alien Species: Observations and Issues from Around the World. 2021;2:16–37.
81. Mehraj G, Khuroo AA, Qureshi S, Muzafar I, Friedman CR, Rashid I. Patterns of alien plant diversity in the urban landscapes of global biodiversity hotspots: a case study from the Himalayas. Biodivers Conserv. 2018;27(5):1055–72.
82. Rashid I, Haq SM, Lembrechts JJ, Khuroo AA, Pauchard A, Dukes JS. Railways redistribute plant species in mountain landscapes. J Appl Ecol. 2021;58(9):1967–80.
83. Gulzar R, Banday FA, Rather ZA, Rashid I, Khuroo AA. Naturalisation of *Ranunculus repens* in Kashmir Himalaya: floristic and ecological aspects. Plant Biosyst Int J Dealing Aspects Plant Biol. 2022;156:1–7.
84. Ricciardi A, Iacarella JC, Aldridge DC, Blackburn TM, Carlton JT, Catford JA, et al. Four priority areas to advance invasion science in the face of rapid environmental change. Environ Rev. 2021;29(2):119–41.
85. McGeoch MA, Spear D, Kleynhans EJ, Marais E. Uncertainty in invasive alien species listing. Ecol Appl. 2012;22(3):959–71.
86. Andersen JC, Wagner DL. Systematics and biological control. In: Van Driesche RG, Simberloff D, Blossey B, Causton C, Hoddle M, Marks CO, Heinz KM, Wagner DL, Warner KD, editors. Integrating biological control into conservation practice. Chichester: Wiley; 2016. p. 105–29.
87. Khuroo AA, Reshi ZA, Rashid I, Dar GH. Towards an integrated research framework and policy agenda on biological invasions in the developing world: a case-study of India. Environ Res. 2011;111(7):999–1006.
88. Albouy V. Amazing invaders: these species from elsewhere. Éditions Quæ, Versailles; 2017. p. 160. ISBN: 978–2-7592-2661-0
89. Fried G. A guide to invasive plants. 2nd ed. Belin, Paris; 2017. p. 304. ISBN: 978-2-410-00417-5

Chapter 2
How to Use the Field Manual?

2.1 Study Region

This *Manual* is based on the research work carried out by the authors in the Kashmir Himalaya (Fig. 2.1). For documentation of the alien flora, we regularly surveyed the study region for field sampling, collection of data and plant specimens from 2019 to 2021. The study region lies between coordinates 33° 20′ to 34° 50′ North Latitude, 73° 55′ to 75° 35′ East Longitude and altitude of the region ranges from 1600 to 5420 m amsl [1]. Kashmir Himalaya covers an area of ~15,000 km² with 64% of the area being mountainous [2]. The 'Karawas', which are plateau-like tablelands developed during the Pleistocene Ice Age and made of clay, sand, and silt of lacustrine origin, are a prominent geological feature of the region [3]. The region, located at the junction of the Holarctic and Paleotropical Floristic Realms, is home to a rich floristic diversity of considerable scientific interest and economic promise [4]. The stunning scenery of this biodiversity-rich area has drawn visitors from far and wide from very early times. This is one of the key reasons for the intentional or unintentional introduction of various floral elements from various phytogeographical regions of the world, which have been supported by distinct bioclimates, a wide range of elevations, and habitat heterogeneity. In fact, the region has long served as a major halting point for historical trade routes travelling from far-east Asia to the Mediterranean coast via Central Asia and vice versa [5]. The climate is primarily of continental temperate type with cold and wet winters and relatively dry and hot summers [6]. The temperature of the region ranges from an average daily maximum of 31°C and minimum of 15°C during summer to an average daily maximum of 4°C and minimum of -4°C during winter and receives an average annual precipitation of ca. 1055 mm, mostly in the form of snow. The natural vegetation of the region mostly consists of alpine meadows and coniferous forests [1].

R. Gulzar et al., *Field Manual on Alien Flora of Kashmir Himalaya*,
Invading Nature - Springer Series in Invasion Ecology 15,
https://doi.org/10.1007/978-3-031-33847-2_2

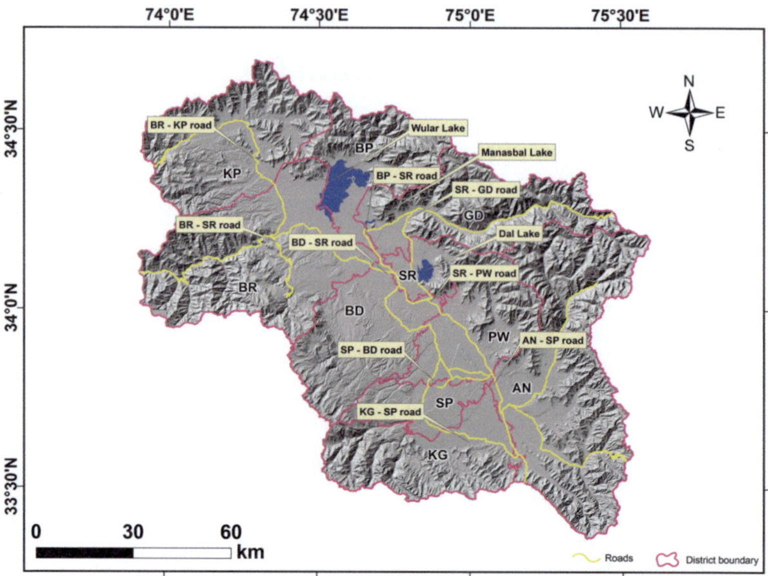

Fig. 2.1 Study area (BR = Baramulla, KP = Kupwara, SR = Srinagar, BP = Bandipore, GD = Ganderbal, BD = Budgam, PW = Pulwama, SP = Shopian, KG = Kulgam, AN = Anantnag)

2.2 Methods Used

This *Manual* is an outcome of the data generated from the field, herbarium and laboratory-based studies. Standard scientific methods were followed for the collection and processing of plant specimens [7]. The place of collection, latitude, longitude, date and time of recording individual species was noted. The diagnostic characters of the plant specimens were photographed in the field and laboratory (by RG[1] & AAK[2]) and included in the *Manual* to supplement information on each species. The plant specimens were identified with the help of relevant taxonomic literature (by RG & AAK). The authentically identified voucher specimens have been deposited in the University of Kashmir Herbarium (KASH). For evaluation of invasion status, we employed the terminology proposed by Pyšek et al. [8]. Geocoordinate data was recorded with an eTrex 30x global positioning system (Garmin, New Taipei City, Taiwan). The coordinates obtained for each species were used for preparing distribution maps using *Arc GIS* Software (version 10.2). For this, we set the appropriate coordinate system in *Arc GIS* using the Universal Tranverse Mercator (UTM) grid option and then selected the zone that applies to the study region. The Shuttle Radar Topography Mission (SRTM) data file was used for each species.

---

1 RG = Ruquia Gulzar.

2 AAK = Anzar Ahmad Khuroo.

Elevation was divided into bands with each band designated with a different colour. Finally, for each species included in the *Manual*, a distribution map showing the locations of occurrence was prepared with reference to elevation zones of the study region.

2.3 Terminology and Definitions

We have followed globally standardised terminology and definitions to present concise scientific information on each alien plant species included in this *Manual*. The alien species' profile information is organised under the following subheadings:

- **Scientific name:** Binomial consisting of two parts: genus followed by species or subspecies or variety epithets and then author name and publication. The *Manual* follows the International Plant Names Index (**IPNI**) [9] for author name and publication of each scientific name.
- **Family:** Family is a taxonomic rank, consisting of one or more genera. For example, Asteraceae (sunflower family) and Brassicaceae (mustard family). The *Manual* follows the Angiosperm Phylogeny Group Classification-IV for family arrangement (APG **IV,** [10]).
- **English name:** For English names, the *Manual* follows Flowers of India [11] and Germplasm Resources Information Network (GRIN) Taxonomy [12]. In case, there were many English names available, the most commonly used has been preferred for each species.
- **Local name:** These are mostly Kashmiri names and are kept in single quotation marks.
- **Habit:** Habit is defined as the general appearance of a plant. On the basis of habit, plants have been categorised into herbs, trees, shrubs and subshrubs. All the species were assigned habit categories based on the field observations following standard terminology [13, 14].

  - **Herb:** A non-woody plant, or one that is woody only at the base.
  - **Tree:** A tree is a woody, perennial plant, usually tall, with a single trunk that bears a crown of branches.
  - **Shrub:** A woody, perennial plant, generally smaller than a tree and with several stems arising from the ground level.
  - **Subshrub**: A low shrub, sometimes with partially herbaceous stems.

- **Life span:** For categorisation of each species into different life spans, terminology proposed by Hickey et al. [13] was used.

  - **Annual:** A plant that completes its life cycle within a single year.
  - **Biennial:** A plant that completes its life cycle within 2 years; producing only vegetative growth in the first year, and flowering in the second.
  - **Perennial:** A plant that lives for a number of years, extending above ground for at least more than 2 years.

For taxonomic description and preparation of photo illustrations of the plant species, we have used freshly collected plant specimens and in some cases herbarium specimens. The ecological traits provided for each species are primarily based on field observations and supplemented with online trait databases (e.g. TRY [15] and BiolFlor [16]).

Taxonomic field characters: For vegetative (leaves) and reproductive parts (flower, fruit) of plant species, we have followed terminology of Singh [17]. In case of leaves, various characters and the terminology used include:

- **Leaf shape**

  - **Cordate:** Heart shaped.
  - **Compound:** Leaf blade divided into two or more blades or segments.
  - **Elliptic:** Broadest at the middle with two equal rounded ends.
  - **Lanceolate:** Narrowly ovate and tapering to a point at the apex.
  - **Linear:** Narrow and much longer than wide, with parallel margins.
  - **Oblong:** Longer than broad, with the margins parallel for most of their length.
  - **Obovate:** Egg-shaped with broadest part near the apex.
  - **Ovate:** Egg-shaped with broadest part near the middle or base.

- **Leaf surface**

  - **Glabrous:** Not covered with any hairs.
  - **Hirsute:** Covered with long stiff hairs.
  - **Hirtellous:** Covered with soft hairs.
  - **Pubescent:** Surface covered soft short hairs.
  - **Strigose:** Covered with stiff appressed hairs pointed in one direction.
  - **Scabrous:** Covered with short rough hairs.
  - **Tomentose:** Covered with matted soft hairs, wooly in appearance.

- **Leaf apex**

  - **Acuminate:** Tapering to a long tip.
  - **Acute:** Pointed tip with sides forming an acute angle.
  - **Mucronate:** Ending abruptly in a short stiff point.
  - **Obtuse:** Broad apex with two sides forming an obtuse angle.
  - **Retuse:** Notched, with a rounded indentation.

- **Leaf base**

  - **Amplexicaul:** Leaf base embracing the stem.
  - **Attenuate:** Showing a long gradual taper towards the base.
  - **Cuneate:** Wedge-shaped base.
  - **Round:** With a broad arch at the base.
  - **Truncate:** Ending abruptly in a more or less straight line, as if cut off.

- **Leaf margin**

  - **Crenate:** Margin notched with regular, rounded symmetrical teeth.
  - **Dentate:** Prominently toothed with acute symmetrical projections pointing outwards.
  - **Entire:** Smooth margin.
  - **Repand:** When the margin is uneven or wavy, with shallow undulations not so deep as for sinuate margins.
  - **Serrate:** Toothed like a saw, with regular acute and angled teeth pointing towards the apex.
  - **Sinuate:** With an uneven margin that has rather deep rounded sinusoidal undulations.

- **Inflorescence:** Arrangement of flowers on a plant is called inflorescence. Some of the common inflorescence types include:

  - **Capitulum:** Characteristic of Asteraceae family. Peduncle is flattened-like head and the florets are attached directly to it.
  - **Corymb:** An indeterminate inflorescence consisting of upper short pedicellate flowers and lower long pedicellate flowers so that all the flowers reach the same level.
  - **Cyme:** It is a determinate inflorescence.
  - **Panicle:** An indeterminate inflorescence, consisting of several branched axes bearing pedicellate flowers.
  - **Raceme:** An indeterminate inflorescence consisting of a single axis bearing pedicellate flowers.
  - **Solitary axillary:** Single flower is present at the axillary position.
  - **Solitary terminal:** One-flowered inflorescence where flower is present at terminal position.
  - **Spike:** A spike is a raceme, but the flowers are sessile.
  - **Synflorescence:** A compact arrangement of capitula within a common involucre.
  - **Umbel:** A determinate or indeterminate, flat-topped or convex inflorescence with pedicels attached at one point.
  - **Verticillaster:** A secondary inflorescence with an indeterminate central axis bearing opposite, lateral, sessile cymes, the flowers appear congested.

- **Fruit type:** Ripened ovary of a flower is called fruit. Some of the common fruit types include:

  - **Achene:** An indehiscent, one-seeded, dry, fruit with seed attached to pericarp at one point only.
  - **Berry:** A fleshy fruit with succulent pericarp.
  - **Capsule:** A dry dehiscent fruit derived from a syncarpous ovary.
  - **Caryopsis:** A small, one-seeded, dry indehiscent fruit with the pericarp adherent to the seed coat.
  - **Cypsela:** A one-seeded fruit, formed from a syncarpous, inferior ovary.

- **Follicle:** A dry, dehiscent fruit derived from one carpel that splits along one suture.
- **Nut:** A one-seeded, dry indehiscent fruit with a hard pericarp usually derived from unilocular ovary.
- **Pod:** A dehiscent fruit, splitting along two sides.
- **Samara:** A single-seeded, dry indehiscent fruit, having a wing-like extension of pericarp.
- **Siliqua:** Fruit developing from bicarpellary syncarpous superior ovary.
- **Utricle:** A thin-walled, one-seeded fruit that dehisces at maturity.

- **Pollination**: The transfer of mature pollen grains to the receptive stigma. Types of pollination on the basis of the agent involved:

  - **Anemophily:** Pollination carried out by wind.
  - **Chiropterophily:** Pollination carried out by bats.
  - **Entomophily:** Pollination carried out by insects.
  - **Hydrophily:** Pollination carried out by water.
  - **Myrmecophily:** Pollination carried out by ants.
  - **Ornithophily:** Pollination carried out by birds.
  - **Zoophily:** Pollination carried out by animals.

- **Seed dispersal**: It is defined as the movement or spread of seeds away from the parent plant.

  - **Autochory:** Seed dispersal by explosion.
  - **Zoochory:** Seed dispersal through animals.
  - **Ornithochory:** Seed dispersal through birds.
  - **Hydrochory:** Seed dispersal through water.
  - **Anemochory:** Seed dispersal through wind.

- **Habitat:** Habitat refers to the place where an organism or population lives or exists.

  - **Agri-fields:** All the land area used for cultivation of crops, e.g. paddy, maize and pulses.
  - **Orchards:** All the land area used for cultivation of horticultural plants, e.g. apple and walnut.
  - **Grasslands:** Land area dominated by herbaceous vegetation, e.g. forbs and grasses. It includes both low-land pastures and high-land meadows.
  - **Roadsides:** Land area close to the roads.
  - **Riparian:** Interface area between land and a waterbody like rivers, lakes, streams and wetlands.
  - **Forests:** Natural land area dominated by trees.
  - **Hillslopes:** Slopes of lower hills.
  - **Gardens:** The land area used for planned cultivation of ornamental plants.

- **Impacts:** The effects of alien plant species on biodiversity, ecology, environment, economy, human health and livestock.

- **Current status:** The current invasion status was assigned according to the standard terminology [8, 18].

  1. **Casual:** Alien plants that may flourish and even reproduce occasionally in an area, but which do not form self-replacing populations, and rely on repeated introductions for their persistence in the area.
  2. **Naturalised:** Alien plants that reproduce consistently and sustain populations over many life cycles without direct intervention by humans but do not necessarily invade natural, seminatural or human-dominated ecosystems.
  3. **Invasive:** Naturalised plants that produce reproductive offspring, often in very large numbers, at considerable distances from parent plants and thus have the potential to spread over a considerable area.

- **Nativity:** Nativity is defined as the region of origin of a plant species. A plant species is said to be native to a particular region if it occurs naturally in an area throughout its evolutionary history. Whilst a plant species that does not occur naturally in a particular area, but has been accidentally or deliberately introduced by humans to the new region is said to be alien (or exotic). The *Manual* follows Germplasm Resource Information Network, Version 2.3.1.1 (https://npgsweb. ars-grin.gov/gringlobal/taxon/taxonomysearch) and Plants of the World Online [19], for designating native range of each plant species.

- **Global distribution:** The global distribution of each plant species includes both the native range and the introduced regions. For this, the *Manual* follows Germplasm Resource Information Network, Version 2.3.1.1 (https://npgsweb. ars-grin.gov/gringlobal/taxon/taxonomysearch). The global distribution differs from the traditional geographical continents in the sense that Asia is divided into two botanical continents (Asia-tropical and Asia-temperate); Australasia includes Australia and New Zealand; majority of the islands in the Pacific Ocean are recognized under the Pacific; and Northern America and Southern America are delimited somewhat differently from the traditional North America and South America.

- **Distribution map:** The map has been prepared in *Arc GIS* 10.2 (ESRI). The differently coloured dots on the map indicate distribution points or locations where the species has been recorded in Kashmir Himalaya. The elevational range for each species has been divided into bands and range of elevation is according to the distribution of each species. The colour of the dot represents a particular elevation band (Fig. 2.2). Most of the species fall under all the five elevation bands because of their wide distribution in the region. Nonetheless, some species, which were recorded from 1 or 2 locations only, were represented in the map by a few dots, falling within a single or two elevation bands. The range of elevation may or may not be the same amongst the species.

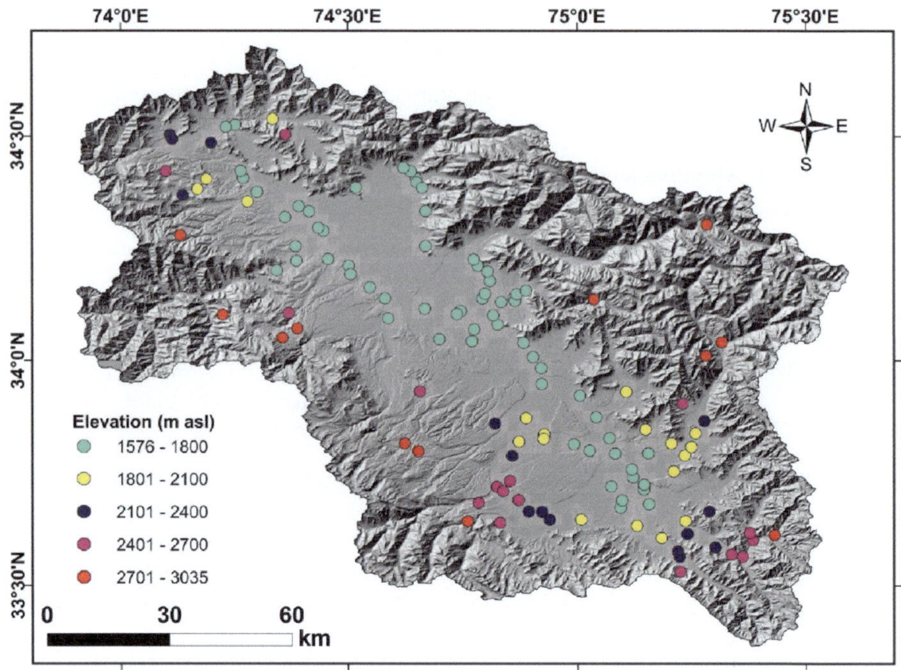

Fig. 2.2 A sample of species distribution map

References

1. Dar GH, Khuroo AA. *Biodiversity of the Himalaya: Jammu and Kashmir state* 263, vol. 18. Singapore: Springer Nature; 2020. p. 1–13.
2. Rashid I, Majeed U, Aneaus S, Cánovas JAB, Stoffel M, Najar NA, Lotus S. Impacts of erratic snowfall on apple orchards in Kashmir Valley, India. Sustainability. 2020;12(21):9206.
3. Kachroo P. Central Asia and Kashmir Himalaya—archaeobotany and floristics. Jodhpur: Scientific Publishers; 1995.
4. Dar GH, Khuroo AA. Floristic diversity in the Kashmir Himalaya: progress, problems and prospects. Sains Malaysiana. 2013;42(10):1377–86.
5. Khuroo AA, Rashid I, Reshi Z, Dar GH, Wafai BA. The alien flora of Kashmir Himalaya. Biol Invasions. 2007;9(3):269–92. https://doi.org/10.1007/s10530-006-9032-6.
6. Gulzar R, Khuroo AA, Rather ZA, Ahmad R, Rashid I. *Symphyotrichum subulatum* (Michx.) GL Nesom (Asteraceae): a new distribution record of an alien plant species in Kashmir Himalaya, India. Check List. 2021;17:569.
7. Bridson D, Forman L. The herbarium handbook. Royal Botanic Gardens, Kew; 1998. p. 291–59.
8. Pyšek P, Richardson DM, Rejmánek M, Webster GL, Williamson M, Kirschner J. Alien plants in checklists and floras: towards better communication between taxonomists and ecologists. Taxon. 2004;53(1):131–43.
9. IPNI. International plant names index. The Royal Botanic Gardens, Kew. Harvard University Herbaria & Libraries and Australian National Botanic Gardens; 2021. Published on the internet http://www.ipni.org.

10. Angiosperm Phylogeny Group, Chase MW, Christenhusz MJ, Fay MF, Byng JW, Judd WS, Soltis DE, Mabberley DJ, Sennikov AN, Soltis PS, Stevens PF. An update of the Angiosperm Phylogeny Group classification for the orders and families of flowering plants: APG IV. Botanical journal of the Linnean. Society. 2016;181(1):1–20.
11. Flowers of India. Available from http://www.flowersofindia.net. Accessed 1 June 2021.
12. GRIN-Germplasm Resources Information Network. Available from https://npgsweb.ars-grin.gov/gringlobal/taxon/taxonomysearch. Accessed 1 June 2021.
13. Hickey M, King C, King M. The Cambridge illustrated glossary of botanical terms. Cambridge University Press; 2000.
14. Mabberley DJ. The plant-book: a portable dictionary of the vascular plants. Cambridge University Press; 1997.
15. Plant Trait Database Available on: https://www.try-db.org/. Accessed on 3 June 2021.
16. Biol Flor. Available from http://www.ufz.de/biolflor. Accessed 30 May 2021.
17. Singh G. Plant systematics: an integrated approach. CRC Press; 2019.
18. Richardson DM, Pyšek P, Rejmánek M, Barbour MG, Panetta FD, West CJ. Naturalization and invasion of alien plants: concepts and definitions. Divers Distrib. 2000;6(2):93–107.
19. POWO. Plants of the World Online. Facilitated by the Royal Botanic Gardens, Kew; 2019. Published on the Internet. http://www.plantsoftheworldonline.org/.

Chapter 3
Alien Species' Profile

R. Gulzar et al., *Field Manual on Alien Flora of Kashmir Himalaya*,
Invading Nature - Springer Series in Invasion Ecology 15,
https://doi.org/10.1007/978-3-031-33847-2_3

Abutilon theophrasti Medik., Malvenfam. 28 (1787).

Family	Malvaceae
English names	Button weed, Velvetleaf
Local name	'Velvat-pan sotchel'
Habit	Annual herb, 100–200 cm tall
Stem	Erect, branched, pubescent, glandular
Leaves	Alternate, petiolate, leaf-blade cordate or ovate, surface velvety, hairy, base cordate, apex acuminate, margin crenate to dentate, stipules caducous
Inflorescence	Solitary axillary or in panicle
Flower	Calyx long, fused below middle, lobes lanceolate, stellate tomentose, corolla of five petals, yellow-orange, staminal tube long, ovary glabrous
Fruit	Capsule
Pollination	Autogamy
Seed dispersal	Autochory
Habitat	Orchards, forests, roadsides, agri-fields
Current status	Invasive
Impacts	Decreases native plant diversity; reported to act as host to different diseases and pests of crops like corn, soyabean and other rain-fed crops
Native range	Africa, Asia-Temperate, Asia-Tropical, Europe
Global distribution	Africa, Asia-Temperate, Asia-Tropical, Europe, Northern America

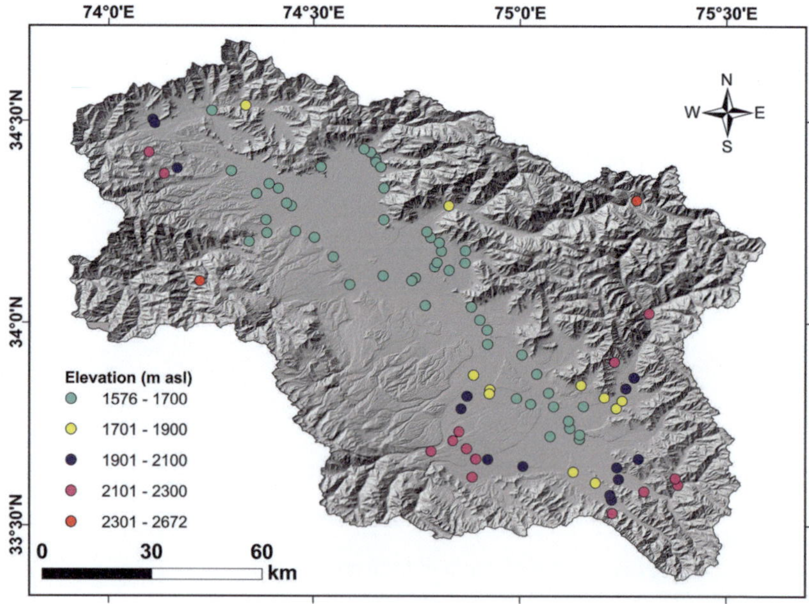

Abutilon theophrasti: (**a**) Habit & Habitat, (**b**) Leaf, (**c**) Flower, (**d**) Fruit

Aesculus indica (Wall. ex Cambess.) Hook.f., Bot. Mag. 85: t. 5117 (1859).

Family	Sapindaceae
English name	Indian horse-chestnut
Local name	'Hani duun'
Habit	Tree, 20–25 m tall
Stem	Erect, branched, girth of trunk up to 1 m
Leaves	Petiolate, palmate, leaflets 5–9, elliptic or lanceolate, base cuneate, apex acuminate, margin serrate
Inflorescence	Panicle
Flower	Bisexual, calyx tube campanulate, petals white with a slight stain of yellow colour
Fruit	Capsule
Pollination	Entomophily
Seed dispersal	Ornithochory
Habitat	Gardens, roadsides, forests
Current status	Naturalised
Impacts	Decreases native plant diversity; fruit reported to be toxic to livestock when consumed in large quantities
Native range	Asia-Tropical, Asia-Temperate
Global distribution	Asia-Tropical, Asia-Temperate

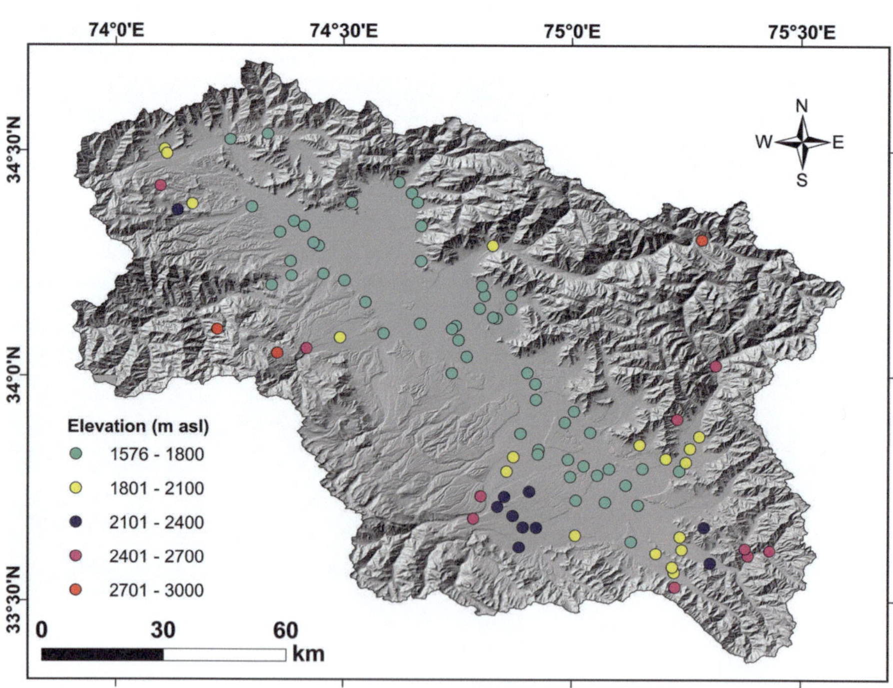

Aesculus indica: (**a**) Habit & Habitat, (**b**) Leaf, (**c**) Inflorescence, (**d**) Fruit

Ageratum conyzoides L., Sp. Pl. 2: 839 (1753).

Family	Asteraceae
English name	Goat weed
Local name	'Tchawul gasi'
Habit	Annual herb, 60–120 cm tall
Stem	Erect, branched, hairy, reddish towards apex
Leaves	Opposite, petiolate, leaf-blade ovate to elliptic, pubescent, margin crenate to serrate
Inflorescence	Capitula small, in dense terminal corymbs, involucre campanulate or hemispheric
Flower	Florets tubular, bisexual, surrounded by 2–3 rows of pointed bracts, margin membranous
Fruit	Achene
Pollination	Entomophily
Seed dispersal	Anemochory, zoochory
Habitat	Gardens, grasslands, roadsides
Current status	Casual
Impacts	Reduces crop yield, allelopathic, decreases native plant diversity, reported to be allergic
Native range	Southern America
Global distribution	Africa, Asia-Temperate, Asia-Tropical, Europe, Northern America, Pacific, Southern America, Australasia

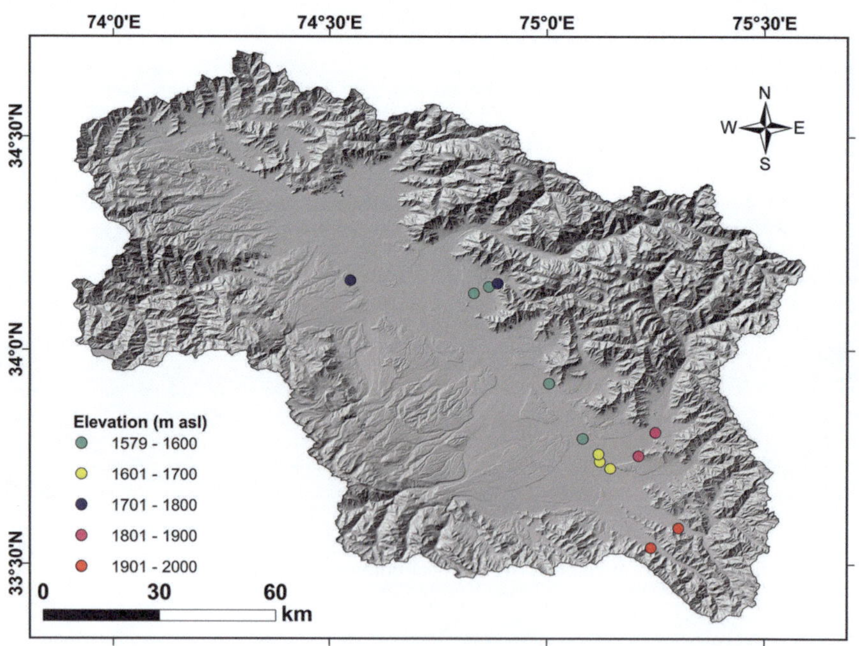

Ageratum conyzoides: (**a**) Habit, (**b**) Leaf, (**c**) Inflorescence, (**d**) Fruit

Ailanthus altissima (Mill.) Swingle, J. Wash. Acad. Sci. 6: 495 (1916).

Family	Simaroubaceae
English name	Tree-of-Heaven
Local name	'Bohr-i kul'
Habit	Tree, 17–20 m tall
Stem	Erect, branched, bark smooth, light grey, often becoming rougher with light tan fissures as the tree ages, ends of branches become pendulous
Leaves	Alternate, odd or even pinnately compound, leaflets ovate, apex acute to acuminate, base cuneate, margin entire
Inflorescence	Large panicles up to 50 cm long on new shoots
Flower	Small, sepals cup-shaped, united, petals yellowish green, valvate, hairy towards inside
Fruit	Samara
Pollination	Entomophily
Seed dispersal	Anemochory
Habitat	Forests, hill slopes, gardens, roadsides, riparian areas
Current status	Invasive
Impacts	Decreases native plant diversity; allelopathic, toxins produced by bark negatively impact human and animal health
Native range	China and Taiwan
Global distribution	Africa, Asia-Temperate, Asia-Tropical, Europe, Northern America, Southern America, Australasia.

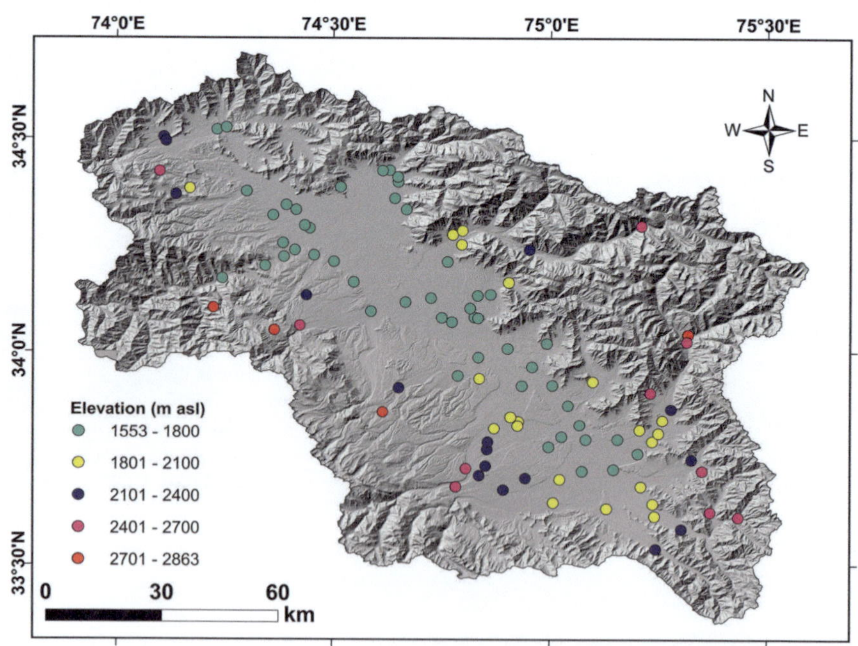

Ailanthus altissima: (**a**) Habit & Habitat, (**b**) Inflorescence, (**c**) Fruit

Alcea rosea L., Sp. Pl. 2: 687 (1753).

Family	Malvaceae
English name	Hollyhock
Local name	'Sazi poosh'
Habit	Annual or perennial herb, 150–300 cm tall
Stem	Erect, rough, hairy
Leaves	Alternate, petiolate, large, leaf-blade ovate, base cordate, margin crenate to dentate, both surfaces pubescent
Inflorescence	Solitary or spike
Flower	Sessile, epicalyx of 6–7 segments, ovate-lanceolate, calyx green, pubescent, corolla of various colours, white, pink, red, yellow
Fruit	Mericarps
Pollination	Entomophily, autogamy
Seed dispersal	Autochory
Habitat	Gardens, grasslands, agri-fields, orchards, roadsides
Current status	Invasive
Impacts	Decreases native plant diversity; declines the forage value of pasturelands
Native range	Asia-Temperate, Europe
Global distribution	Asia-Temperate, Asia-Tropical, Europe, Northern America, Southern America

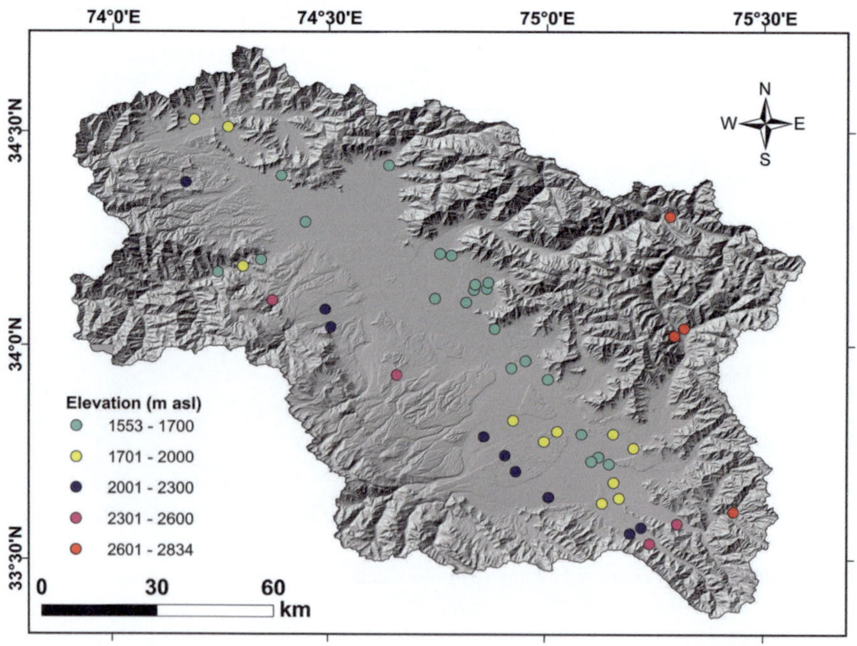

Alcea rosea: (**a**) Habit & Habitat, (**b**) Leaf (**c, d**) Flower, (**e**) Fruit

Alternanthera philoxeroides (Mart.) Griseb., Abh. Königl. Ges. Wiss. Göttingen. 24: 36 (1879).

Family	Amaranthaceae
English name	Alligator weed
Local name	'Magarmachh gasi'
Habit	Perennial herb, 20–120 cm tall
Stem	Ascending or prostrate, branched, hairy when young and glabrous at maturity
Leaves	Opposite, petiolate, leaf-blade ovate to lanceolate, margin entire
Inflorescence	White heads in upper leaf axils, peduncled, globose
Flower	Bisexual, tepals white, shiny, oblong, staminodes linear, ovary compressed
Fruit	Utricle
Pollination	Autogamy
Seed dispersal	Hydrochory
Habitat	Lakes, streams, canals, wetlands
Current status	Naturalised
Impacts	Affects water quality and flow, blocks navigation, reduces light and oxygen levels below the water surface and negatively affects aquaculture; decreases native plant diversity
Native range	Southern America
Global distribution	Asia-Temperate, Australasia, Northern America, Southern America

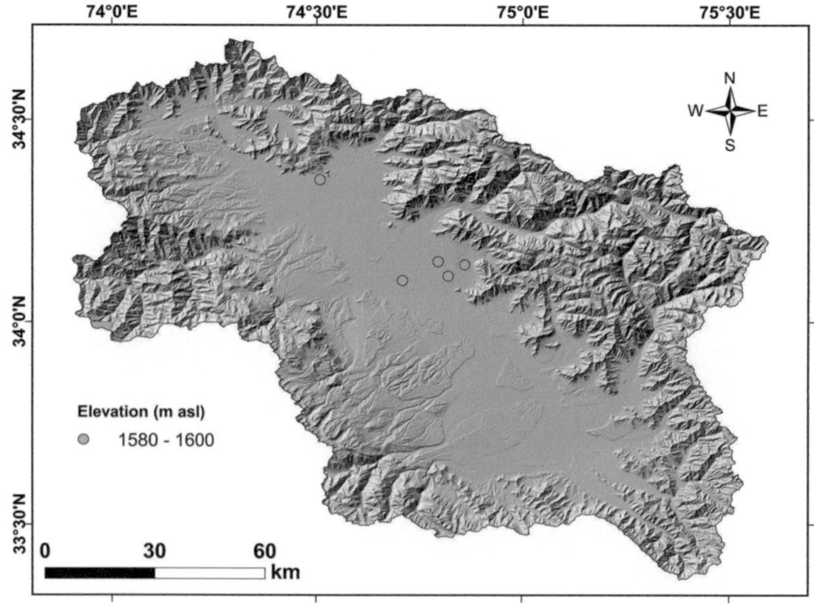

Alternanthera philoxeroides: (**a**) Habit & Habitat (Inset: Flower), (**b**) Leaf, (**c**) Inflorescence (Flowers)

Amaranthus caudatus L., Sp. Pl. 2: 990 (1753).

Family	Amaranthaceae
English names	Tassel flower, Fox-tail amaranth
Local name	'Ganihaar'
Habit	Annual herb, upto 100 cm tall.
Stem	Erect, branched, red or purple, sparsely hairy
Leaves	Alternate, petiolate, leaf-blade ovate to lanceolate, tip acute, margin entire
Inflorescence	Panicle, pendulous
Flower	Perianth segments 5, lanceolate or ovate in male flowers, in female flowers spathulate, stamens 5, stigmas 3
Fruit	Utricle
Pollination	Anemophily
Seed dispersal	Anemochory, zoochory
Habitat	Agri-fields, roadsides, gardens, grasslands, orchards
Current status	Invasive
Impacts	Reduces crop yield; decreases native plant diversity
Native range	Asia-Tropical, Northern America
Global distribution	Africa, Asia-Temperate, Asia-Tropical, Northern America, Southern America

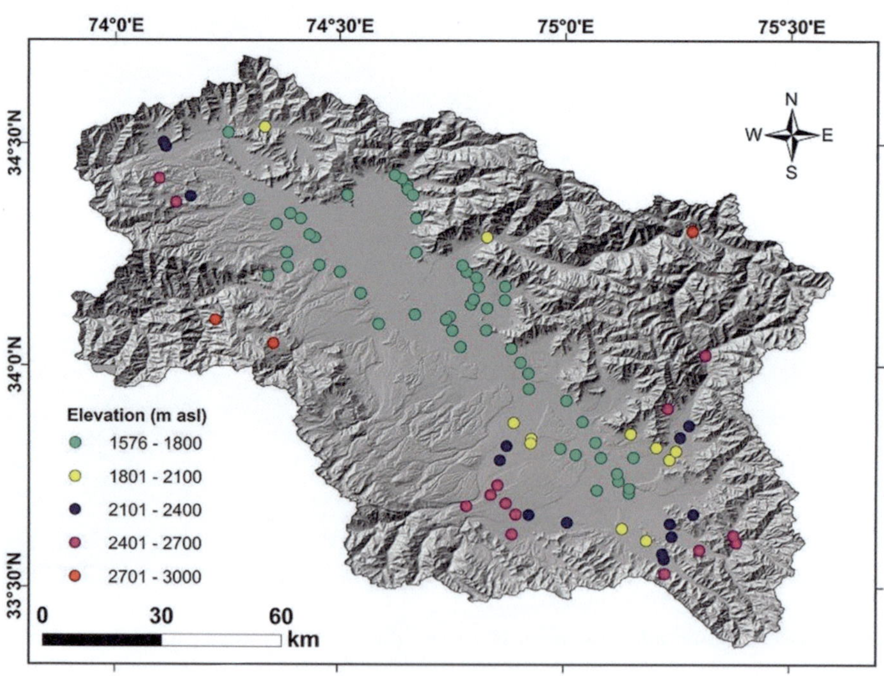

Amaranthus caudatus: (**a**) Habit & Habitat, (**b**, **c**) Inflorescence, (**d**) Perianth, (**e**) Seeds

Amaranthus spinosus L., Sp. Pl. 2: 991 (1753).

Family	Amaranthaceae
English names	Thorny amaranth, Prickly amaranth
Local name	'Kond ganihaar'
Habit	Annual herb, 30–150 cm tall
Stem	Erect, branched, glabrous or sparsely hairy, sometimes reddish
Leaves	Alternate, petiolate, leaf-blade ovate or lanceolate, tip retuse or blunt, margin entire, a pair of fine and compressed spines at each leaf axil
Inflorescence	Spike
Flower	Pistillate flowers with 5 tepals, ovate to lanceolate in shape, bracts deltoid–ovate, stigmas 2–3, staminate flowers terminal, globose, tepals 5, stamens 5
Fruit	Utricle
Pollination	Anemophily
Seed dispersal	Anemochory
Habitat	Grasslands, roadsides, orchards, agri-fields
Current status	Invasive
Impacts	Decreases native plant diversity; reduces crop yield, unpalatable to livestock and thus reduces the forage value of pasturelands
Native range	Northern America, Southern America
Global distribution	Africa, Asia-Temperate, Asia-Tropical, Northern America, Southern America, Australasia

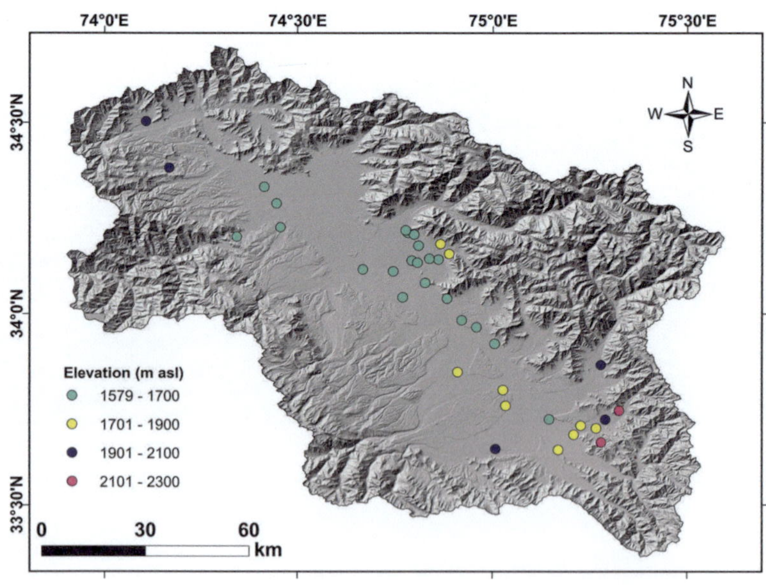

Amaranthus spinosus: (**a**, **b**) Habit & Habitat, (**c**, **d**) Leaf & Inflorescence

Amaranthus viridis L., Sp. Pl., ed. 2. 2: 1405 (1763).

Family	Amaranthaceae
English names	Green amaranth, Slender amaranth
Local name	'Sabz ganihaar'
Habit	Annual herb, 10–100 cm tall
Stem	Erect, slender, sparsely branched, glabrous, hairs only on inflorescence
Leaves	Alternate, petiolate, leaf-blade deltoid or oblong, base cuneate, margin entire, tip obtuse or slightly mucronate
Inflorescence	Paniculate spikes, bracts deltoid or lanceolate
Flower	Green, perianth segments 3 (rarely 4), ovate to oblong in male flower, narrowly spathulate in female flower, stamens 5, stigmas 2–3, short or erect
Fruit	Utricle
Pollination	Anemophily
Seed dispersal	Anemochory, zoochory
Habitat	Agri-fields, gardens, grasslands, orchards, roadsides
Current status	Invasive
Impacts	Declines agricultural productivity; decreases native plant diversity, reported to be poisonous to some animals
Native range	Africa, Asia-Tropical, Southern America
Global distribution	Africa, Asia-Tropical, Southern America, widely naturalised in tropics

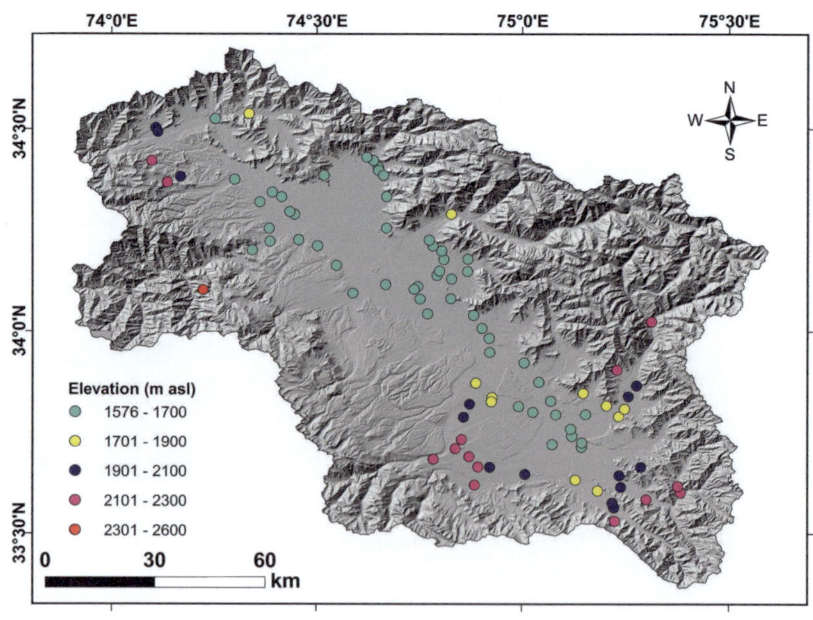

Amaranthus viridis: (**a**) Habit & Habitat, (**b**, **c**) Leaf, (**d**) Inflorescence

Anethum graveolens L., Sp. Pl. 1: 263 (1753).

Family	Apiaceae
English name	Dill
Local name	'Panjeab beedyaani'
Habit	Annual herb, 100–200 cm tall
Stem	Erect, branched, striate, glabrous
Leaves	Petiolate, basal leaf-blade ovate, pinnately divided into linear segments, stem leaves less divided than basal leaves, petioles sheathing
Inflorescence	Umbel
Flower	Sepals inconspicuous, petals yellow
Fruit	Schizocarp
Pollination	Entomophily, autogamy
Seed dispersal	Autochory
Habitat	Gardens, roadsides, orchards, agri-fields
Current status	Invasive
Impacts	Decreases native plant diversity; reduces the quantity and quality of crop yield
Native range	Asia-Temperate, Africa
Global distribution	Africa, Asia-Tropical, Asia-Temperate, Europe, Northern America, Southern America, Australasia

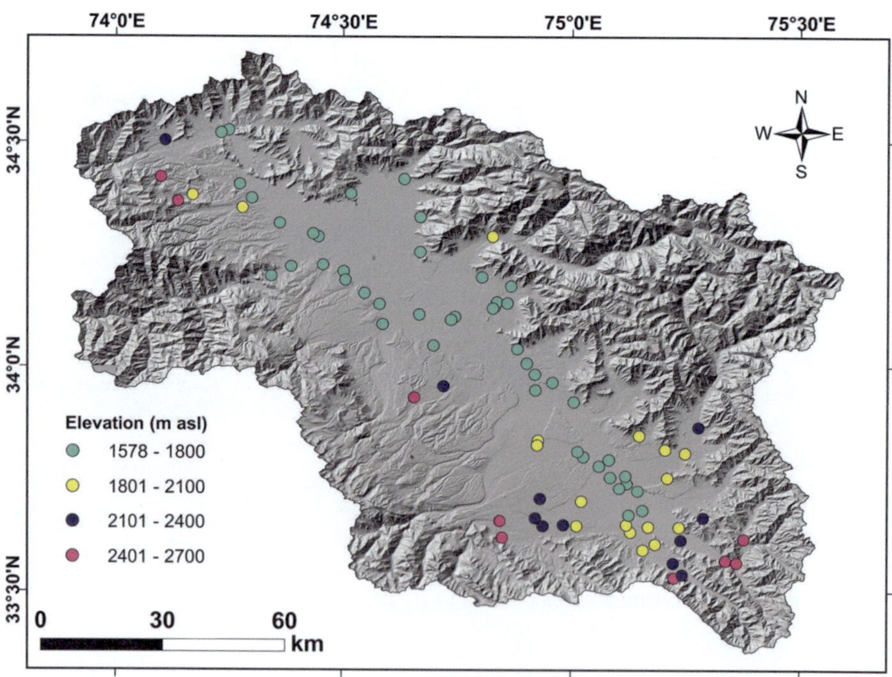

Anethum graveolens: (**a**, **c**) Habit & Habitat, (**b**) Inflorescence, (**d**) Fruit

Anthemis cotula L., Sp. Pl. 894 (1753).

Family	Asteraceae
English names	Stinking chamomile, Dog-fennel
Local name	'Faki gasi'
Habit	Annual herb, up to 60–70 cm tall
Stem	Erect, branched, sparsely hairy
Leaves	Alternate, subsessile or shortly petiolate, leaf-blade obovate-oblong or ovate-oblong, dissected into 2–3-pinnatisect linear segments
Inflorescence	Capitula, solitary, radiate, sometimes discoid, phyllaries ovate to oblong
Flower	Ray florets sterile, tube glabrous, ligules white, disc florets yellow, corolla tube terete, glabrous, slightly inflated at the base
Fruit	Achene
Pollination	Entomophily
Seed dispersal	Anemophily
Habitat	Gardens, orchards, roadsides, riparian areas, grasslands, forests
Current status	Invasive
Impacts	Decreases native plant diversity; declines agricultural productivity, unpalatable to livestock, drastically reduces forage value of pasturelands; causes skin irritation in humans
Native range	Africa, Asia-Temperate, Asia-Tropical, Europe
Global distribution	Africa, Asia-Temperate, Asia-Tropical, Europe, Northern America, Southern America, Australasia

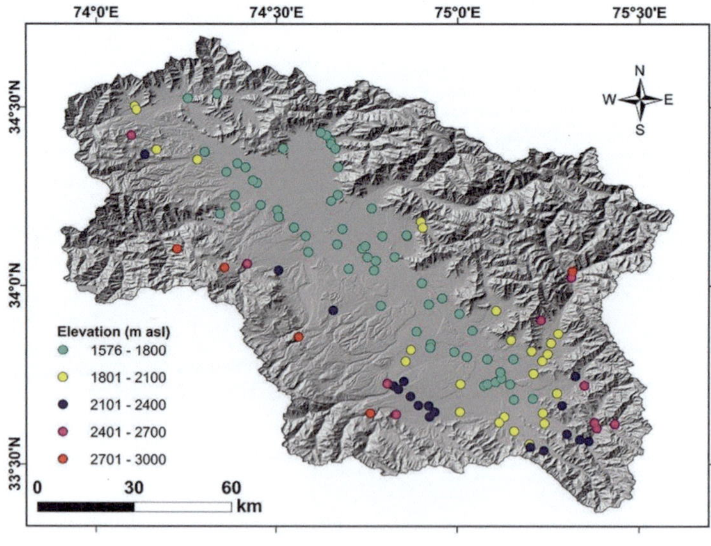

Anthemis cotula: (**a**) Habitat, (**b**) Habit, (**c**) Leaf, (**d**) Inflorescence (Flowers)

Arctium lappa L., Sp. Pl. 2: 816 (1753).

Family	Asteraceae
English name	Greater burdock
Local name	'Tyier-e kul'
Habit	Biennial herb, upto 200 cm tall
Stem	Erect, apically branched, purplish
Leaves	Alternate, petiolate, leaf-blade broadly cordate, margin entire sometimes repand to denticulate, adaxially green and abaxially greyish white, sparsely strigose; basal leaves with long petiole ca. 30 cm; cauline leaves similar to basal leaves or shallowly cordate to ovate
Inflorescence	Capitula, corymbose; involucre globular, phyllaries with hooked apex
Flower	Corolla purplish red
Fruit	Achene (bur)
Pollination	Entomophily, autogamy
Seed dispersal	Zoochory
Habitat	Agri-fields, roadsides, orchards, grasslands, forests
Current status	Invasive
Impacts	Reduces native plant diversity; declines agricultural productivity; decreases grazing potential of pasturelands; reported to cause dermatitis and respiratory problems in humans; spiny fruits (burs) often stuck to skin of livestock and human clothes and cause allergy
Native range	Asia-Temperate, Asia-Tropical, Europe
Global distribution	Asia-Temperate, Asia-Tropical, Europe, Northern America, Australasia

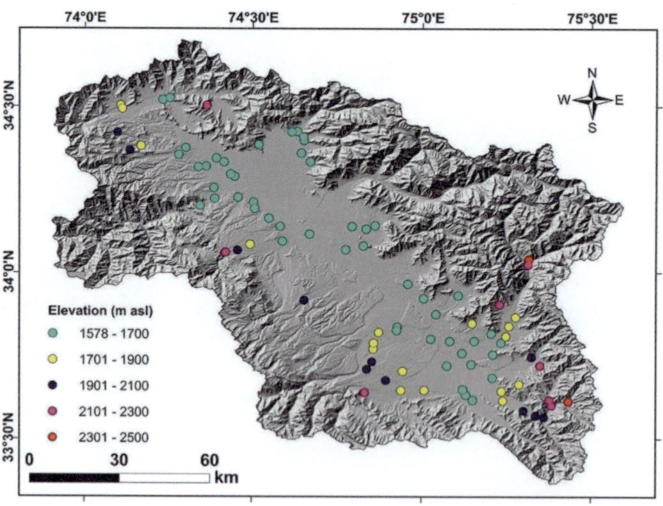

Arctium lappa: (**a**) Habit, (**b**) Stem, (**c**) Leaf, (**d, e**) Inflorescence, (**f**) Florets

Artemisia annua L., Sp.Pl. 2: 847 (1753).

Family	Asteraceae
English names	Sweet wormwood, Annual mugwort
Local name	'Teathwean'
Habit	Annual herb, upto 200 cm tall
Stem	Erect, branched, sparsely pubescent to glabrous, strongly aromatic
Leaves	Basal and lower stem leaves petiolate, leaf-blade ovate to triangular, tri-pinnatisect, margin serrate, middle and upper leaves subsessile to sessile, bi-pinnatisect, upper-most leaves lanceolate, margin entire to toothed
Inflorescence	Capitula, involucre hemispherical, phyllaries 3-seriate
Flower	Disc florets and outer florets pistillate, inner florets bisexual, corolla greenish or yellowish
Fruit	Cypsela
Pollination	Anemophily
Seed dispersal	Autochory
Habitat	Grasslands, roadsides, orchards, forests
Current status	Invasive
Impacts	Decreases native plant diversity; allelopathic, reduces the yield of crops, reported to retard the seed germination of maize
Native range	Asia-Temperate, Europe
Global distribution	Asia-Tropical, Asia-Temperate, Europe, Northern America, Southern America, Australasia

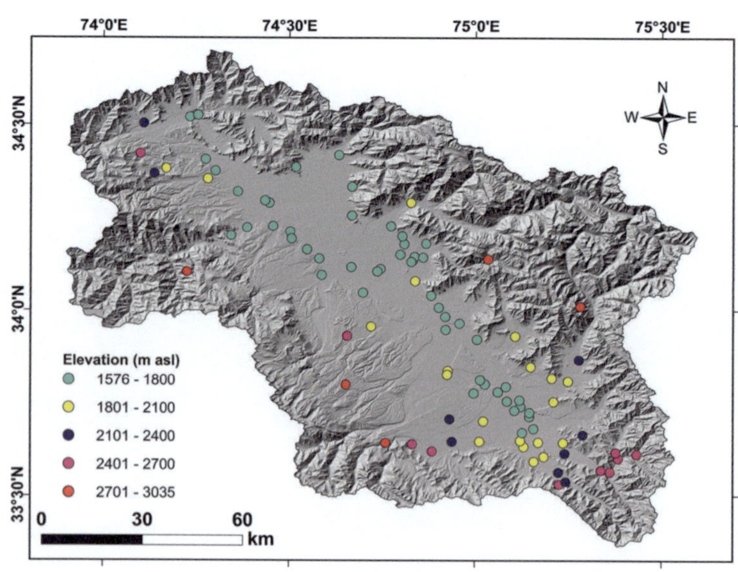

Artemisia annua: (**a**) Habitat (Inset: Leaf), (**b**) Habit, (**c**) Inflorescence

Asparagus officinalis L., Sp. Pl. 1: 313 (1753).

Family	Asparagaceae
English name	'Garden asparagus'
Local name	'Parglass gassi'

Habit	Perennial herb, upto 200 cm long
Stem	Erect, branched, glabrous
Leaves	Reduced to scales, green cladodes serve the function of leaves
Inflorescence	Solitary or in clusters
Flower	Pedicellate, bell-shaped with 6 tepals, yellowish
Fruit	Berries, red in colour
Pollination	Anemophily
Seed dispersal	Ornithochory, zoochory
Habitat	Gardens, grasslands, forests

Current status	Naturalised
Impacts	Alters the nutrient composition of soil, reduces native plant diversity

Native range	Africa, Asia-Temperate, Europe
Global distribution	Africa, Asia-Temperate, Asia-Tropical, Europe, Northern America, Southern America, Australasia

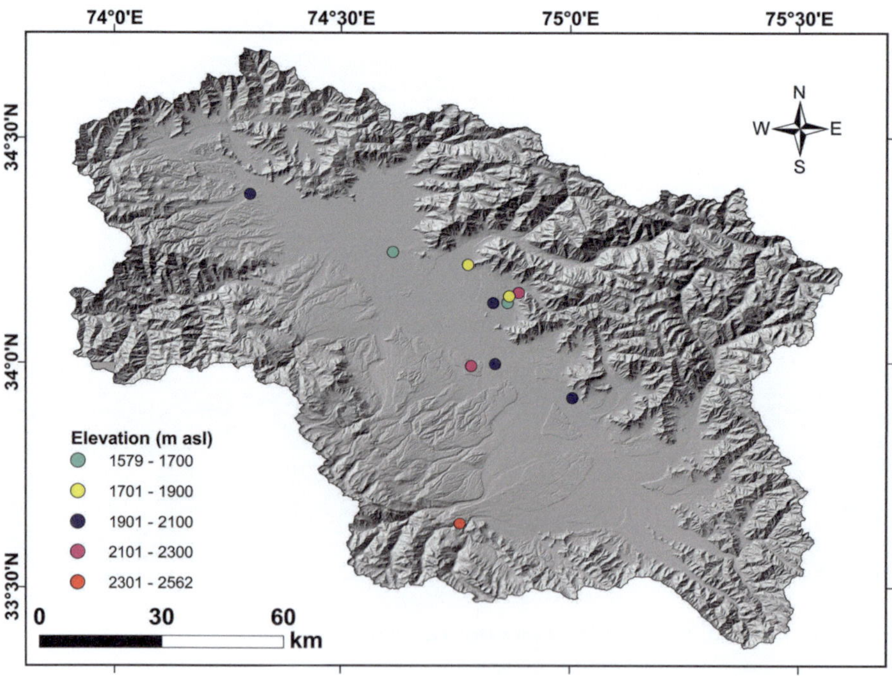

***Asparagus officinalis*:** (**a**) Habit, (**b**) Inflorescence (Flowers), (**c**) Fruit

Bellis perennis L., Sp. Pl. 2: 886 (1753).

Family	Asteraceae
English name	English daisy
Local name	'Meadean dazy'

Habit	Annual or perennial herb
Stem	Scapose, 10–25 cm tall, sparsely hairy
Leaves	Rosulate, evergreen, petiole long, winged, leaf-blade spathulate, base attenuate, tip obtuse, margin serrate to crenate
Inflorescence	Capitula solitary, terminal, involucre hemispherical, phyllaries 2-seriate
Flower	Ray florets white or pinkish, disc florets yellow, campanulate
Fruit	Achene
Pollination	Entomophily
Seed dispersal	Anemochory, zoochory, myrmecochory
Habitat	Forests, gardens, grasslands, riparian areas

Current status	Naturalised
Impacts	Decreases native plant diversity; reported to produce acrid acid in its foliage which makes it unpalatable for insects.

Native range	Africa, Asia-Temperate, Europe
Global distribution	Africa, Asia-Temperate, Europe, Northern America, Southern America Pacific, Australasia

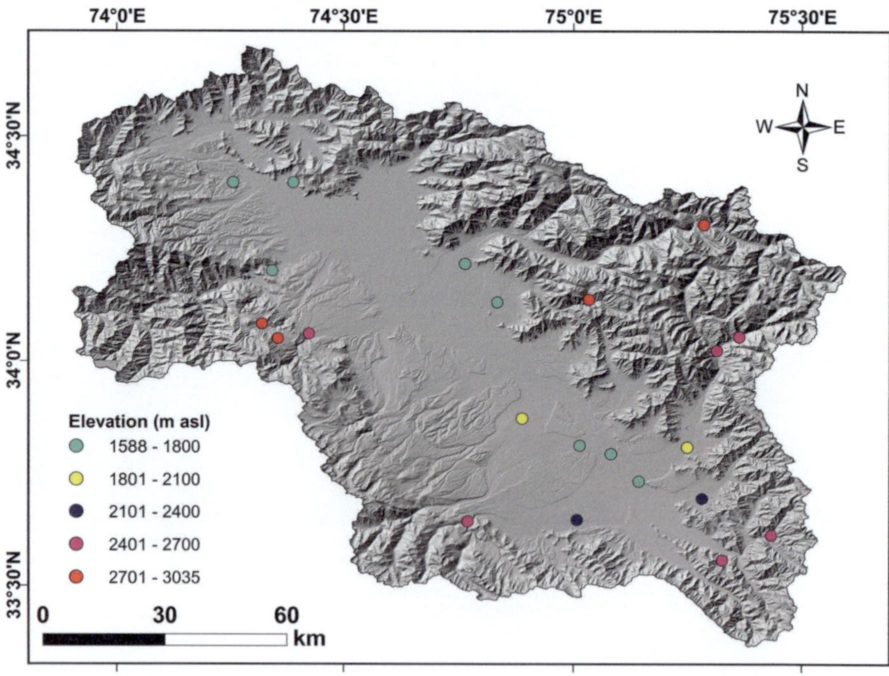

Bellis perennis: (**a**) Habitat, (**b**) Habit, (**c**) Leaf, (**d, e**) Flower (Inflorescence)

Bidens bipinnata L., Sp. Pl. 2: 832 (1753).

Family	Asteraceae
English name	Spanish needles
Local name	'Panjeab si-tsen kul'
Habit	Annual herb, 30–150 cm tall
Stem	Erect, green, veined, glabrous
Leaves	Opposite, petiolate, compound, leaf segments ovate-lanceolate, margin serrate, both surfaces glabrous
Inflorescence	Capitula, phyllaries linear to lanceolate, 2-seriate
Flower	Ray florets petal-like, 3–5, yellowish, disc florets tubular, corolla golden yellow
Fruit	Achene
Pollination	Entomophily
Seed dispersal	Ornithochory, zoochory
Habitat	Grasslands, gardens, roadsides, riparian areas
Current status	Naturalised
Impacts	Reduces yield of many crops; decreases native plant diversity, acts as a host to many pests and diseases
Native status	Northern America
Global distribution	Africa, Asia-Temperate, Asia-Tropical, Europe, Southern America, Australasia

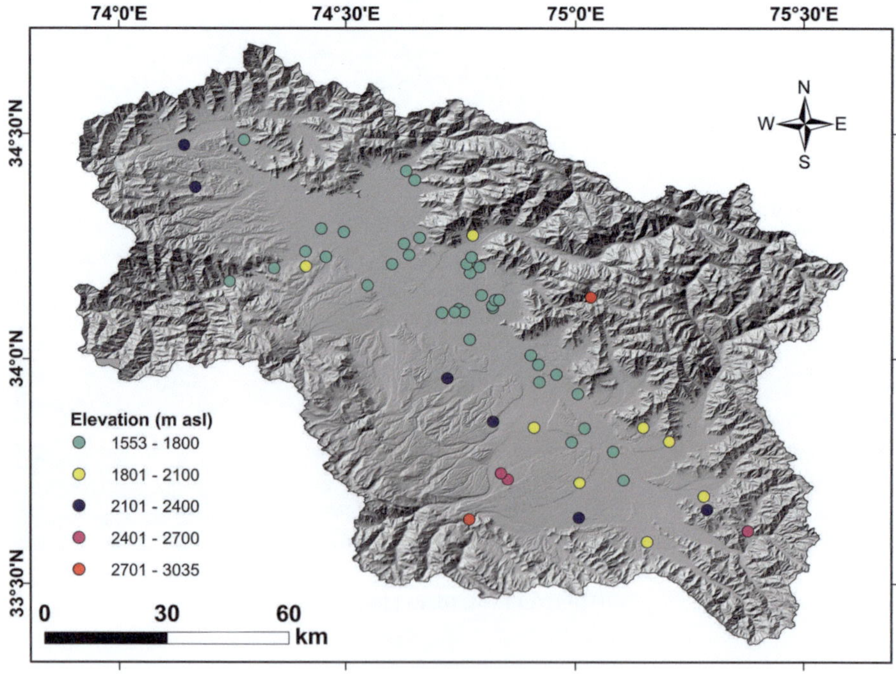

***Bidens bipinnata*:** (**a**) Habit & Habitat, (**b**) Leaf, (**c**) Flower (Inflorescence), (**d**) Fruit

Bromus catharticus Vahl, Symb. Bot. (Vahl) 2: 22 (1791).

Family	Poaceae
English name	Rescue grass
Local name	'Lat-catchal gassi'

Habit	Perennial herb, upto 100 cm tall
Stem	Culms, erect or ascending, unbranched with pigmented hairless nodes
Leaves	Alternate, upto 30 cm long, leaf-blade linear, hairy on upper surface, base attenuate
Inflorescence	Panicle, spikelets 6–12 flowered, compressed
Flower	Glumes long, pointed, keeled, lemmas lanceolate, awn short or absent
Fruit	Caryopsis
Pollination	Anemophily
Seed dispersal	Anemochory, zoochory
Habitat	Grasslands, agri-fields, roadsides, gardens, riparian areas

Current status	Invasive
Impacts	Decreases native plant diversity; reduces crop yield, causes damage to hides of animals, reported to act as host of some cereal crop diseases like ergot; seeds reported to cause injury to intestines, if eaten by animals

Native range	Southern America
Global distribution	Asia-Tropical, Southern America

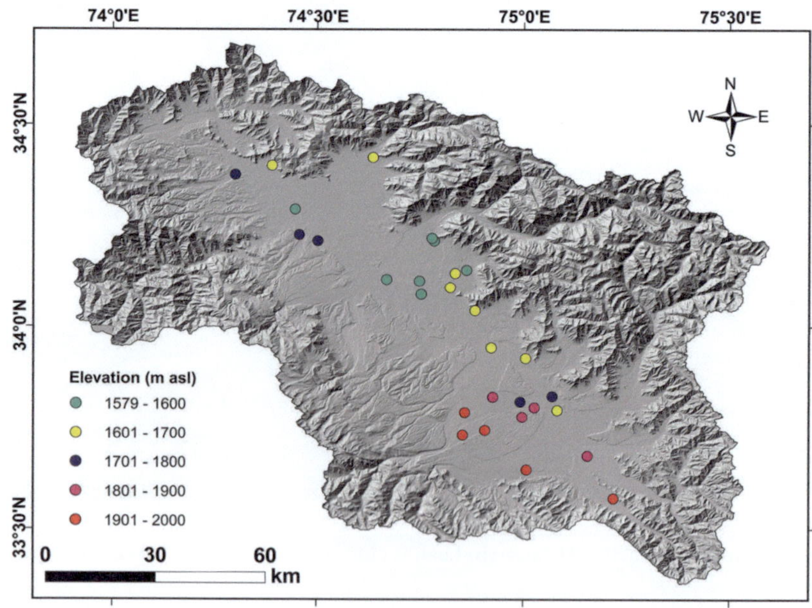

***Bromus catharticus*:** (**a**) Habit & Habitat, (**b**) Sheathing leaf, (**c**) Inflorescence, (**d**) Spikelet

Buddleja davidii Franch., Nouv. Arch. Mus. Hist. Nat. ser. 2, 10: 65 (1888).

Family	Scrophulariaceae
English name	Royal red butterfly-bush
Local name	'Pompur zaraend'
Habit	Deciduous shrub, 5–8 m tall
Stem	Erect, 4-sided, branched
Leaves	Opposite, subsessile, leaf-blade ovate-lanceolate to lanceolate, upper surface dark green, glabrous, lower surface whitish, tomentose, margin serrate, stipules interpetiolar
Inflorescence	Terminal, long panicle
Flower	Bisexual, calyx narrowly campanulate, corolla fused upto three quarters, lilac to purple
Fruit	Capsule
Pollination	Entomophily
Seed dispersal	Anemochory, zoochory
Habitat	Forests, roadsides, gardens, riparian areas
Current status	Naturalised
Impacts	Reduces native vegetation of riparian ecosystems; hinders water flow
Native range	Asia-Temperate
Global distribution	Asia-Temperate, Europe, Northern America, Australasia

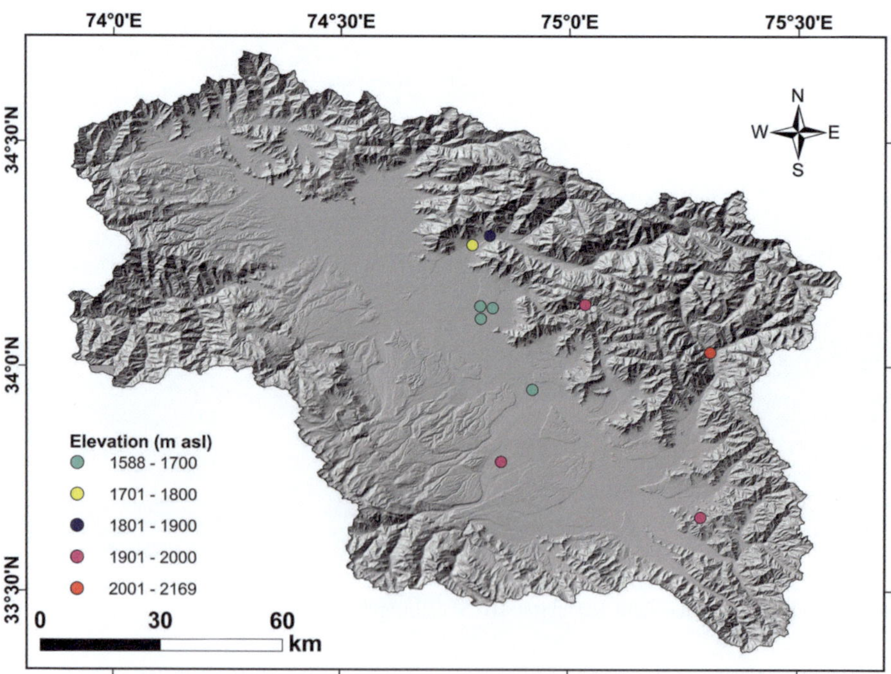

Buddleja davidii: (**a**) Habitat, (**b**) Habit, (**c**) Leaf, (**d**) Inflorescence

Calendula officinalis L., Sp. Pl. 2: 291 (1753).

Family	Asteraceae
English name	Pot marigold
Local name	'Hamyiash bahaar'
Habit	Annual herb, 20–80 cm tall
Stem	Erect, branched, pubescent, green
Leaves	Basal leaves oblong-spatulate, margin serrate or entire, stem leaves sessile, leaf-blade oblong to lanceolate, base amplexicaul, apex obtuse, margin serrate
Inflorescence	Capitula, involucre hemispherical, phyllaries lanceolate, pubescent
Flower	Ray florets yellow or orange, disc florets yellow
Fruit	Achene
Pollination	Entomophily
Seed dispersal	Anemochory, zoochory
Habitat	Gardens, roadsides, agri-fields
Current status	Casual
Impacts	Weed of agriculture, highly competitive, produces a large number of viable seeds
Native range	Europe
Global distribution	Africa, Asia-Temperate, Asia-Tropical, Europe, Northern America, Southern America, Australasia

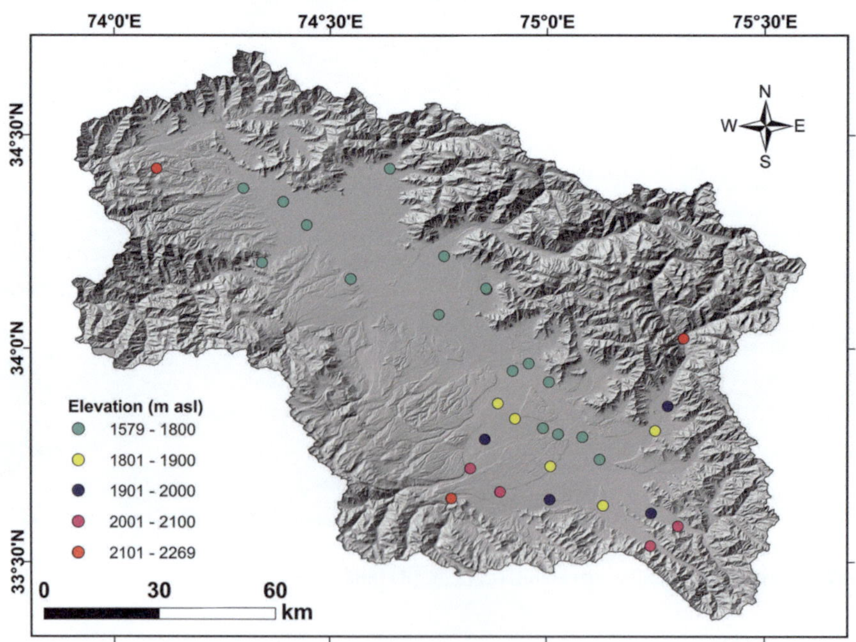

Calendula officinalis: (**a**) Habit, (**b**) Leaf, (**c**) Floral bud, (**d**) Phyllaries, (**e**) Flower (Inflorescence), (**f**) Fruit

Carthamus lanatus L., Sp.Pl. 2: 830 (1753).

Family	Asteraceae
English names	Saffron thistle, Downy safflower
Local name	'Kong-i tchier'
Habit	Annual herb, 40–200 cm tall
Stem	Erect, branched, straw coloured.
Leaves	Basal leaves petiolate, leaf-blade pinnately divided into linear or lanceolate spine-tipped lobes, cauline leaves sessile, leaf-blade lanceolate to ovate, margin with spine-tipped lobes, base clasping, tip spinose, acuminate
Inflorescence	Capitula, involucre ovoid, hairy, phyllaries linear to lanceolate, spine tipped
Flower	Corolla yellow, red or black-veined
Fruit	Cypsela
Pollination	Entomophily
Seed dispersal	Anemochory
Habitat	Orchards, roadsides, grasslands, hill slopes
Current status	Invasive
Impacts	Reduces forage value of pasture lands; decreases native plant diversity; causes injury to eyes and mouth of animals, the spines damage clothes and skin of humans
Native range	Africa, Asia-Temperate, Asia-Tropical, Europe
Global distribution	Africa, Asia-Temperate, Asia-Tropical, Europe, Australasia, Northern America, Southern America

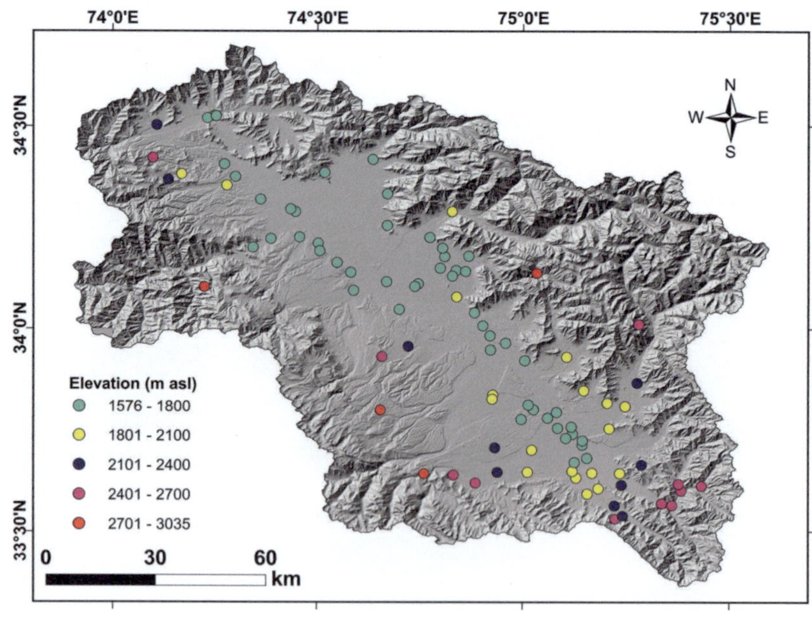

Carthamus lanatus: (**a**, **b**) Habit & Habitat, (**c**, **d**) Inflorescence

Centaurea cyanus L., Sp. Pl. 2: 911 (1753).

Family	Asteraceae
English names	Cornflower, Bachelor's button
Local name	'Tchi-battan posh'
Habit	Annual herb, 30–90 cm tall
Stem	Erect, openly branched, tomentose
Leaves	Basal leaves linear to lanceolate, grey, tomentose, margin entire rarely lobed, cauline leaves alternate, leaf-blade linear, margin entire
Inflorescence	Capitula, involucre ovoid or bell-shaped, phyllaries green to purple, tip fringed with slender teeth
Flower	Florets blue or purple, sterile florets ray like
Fruit	Cypsela
Pollination	Entomophily
Seed dispersal	Autochory, myrmecochory
Habitat	Gardens, roadsides, forests
Current status	Casual
Native range	Asia-Temperate, Europe
Global distribution	Africa, Asia-Temperate, Asia-Tropical, Europe, Northern America, Southern America, Australasia

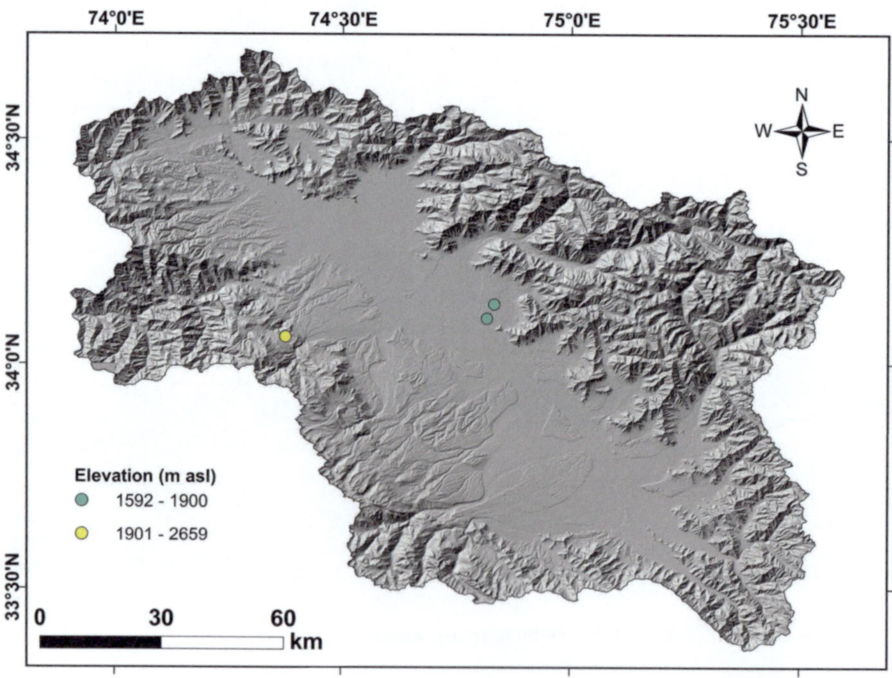

Centaurea cyanus: (**a**, **b**) Habit & Habitat, (**c**) Leaf, (**d**) Floral bud, (**e**) Flower, (**f**, **g**) Fruit

Chenopodium album L., Sp. Pl. 1: 219 (1753).

Family	Amaranthaceae
English names	Fat hen, White goosefoot
Local name	'Lyisi'
Habit	Annual herb, 15–180 cm tall
Stem	Erect, branched, green or with purple streaks
Leaves	Petiolate, leaf-blade rhombic, ovate to lanceolate, base cuneate, apex acute, margin serrate
Inflorescence	Glomerules in panicles or spike-like panicles
Flower	Bisexual, perianth segments five, elliptic to ovate
Fruit	Utricle
Pollination	Anemophily
Seed dispersal	Anemochory, zoochory
Habitat	Agri-fields, roadsides, gardens, orchards, grasslands, forests
Current status	Invasive
Impacts	Decreases native plant diversity; reduces quality and quantity of crop yield, causes economic loss in agriculture, allelopathic, toxic to animals, harmful to dairy cows
Native range	Africa, Asia-Temperate, Europe
Global distribution	Africa, Asia-Temperate, Asia-Tropical, Europe, Northern America, Pacific, Southern America, Australasia

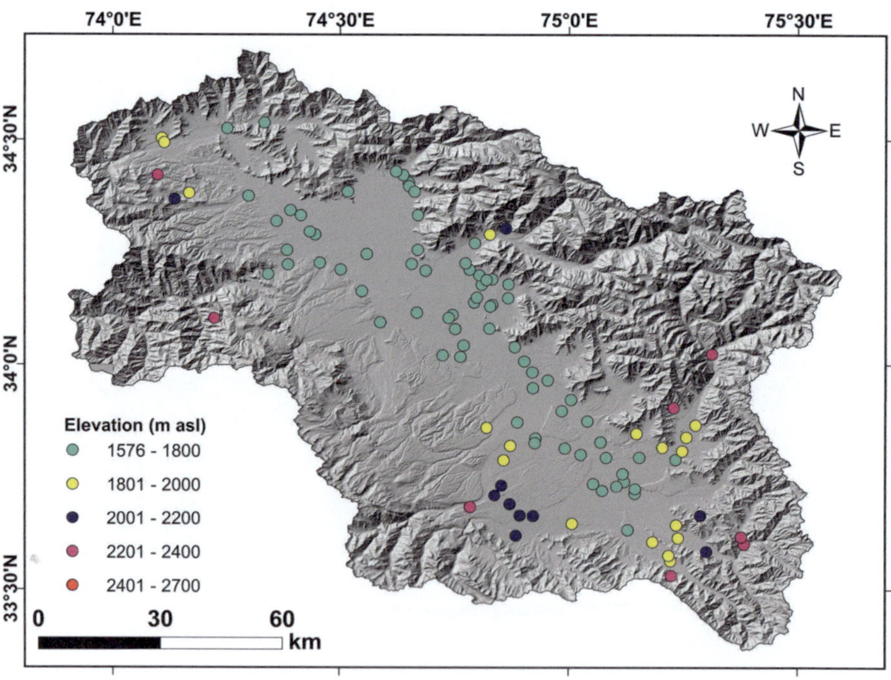

Chenopodium album: (**a**, **b**) Habit & Habitat, (**c**) Stem, (**d**) Leaf, (**e**) Inflorescence, (**f**) Fruit

Cirsium arvense Scop., Fl. Carniol., ed. 2. 2:126 (1772).

Family	Asteraceae
English name	Creeping thistle
Local name	'Kiend kul'

Habit	Perennial herb, 30–180 cm tall
Stem	Stems erect, branched, pubescent
Leaves	Sessile, spiny, lobed, margin coarsely dentate
Inflorescence	Capitula numerous, terminal, corymbose, involucre hemispherical to ovoid, phyllaries 5–8-seriate, imbricate
Flower	All the florets of similar form, corolla purple or pink
Fruit	Cypsela
Pollination	Entomophily
Seed dispersal	Anemochory, zoochory, hydrochory
Habitat	Grasslands, roadsides, gardens, orchards, agri-fields, hill slopes, forests

Current status	Invasive
Impacts	Causes economic loss in agriculture and horticulture; decreases native plant diversity, reduces quality and quantity of crop yield, allelopathic, reduces the forage value of pastures, reduces the aesthetic and recreational value of tourist spots

Native range	Asia-Temperate, Asia-Tropical, Europe
Global distribution	Asia-Temperate, Asia-Tropical, Africa, Europe, Northern America, Southern America, Australasia

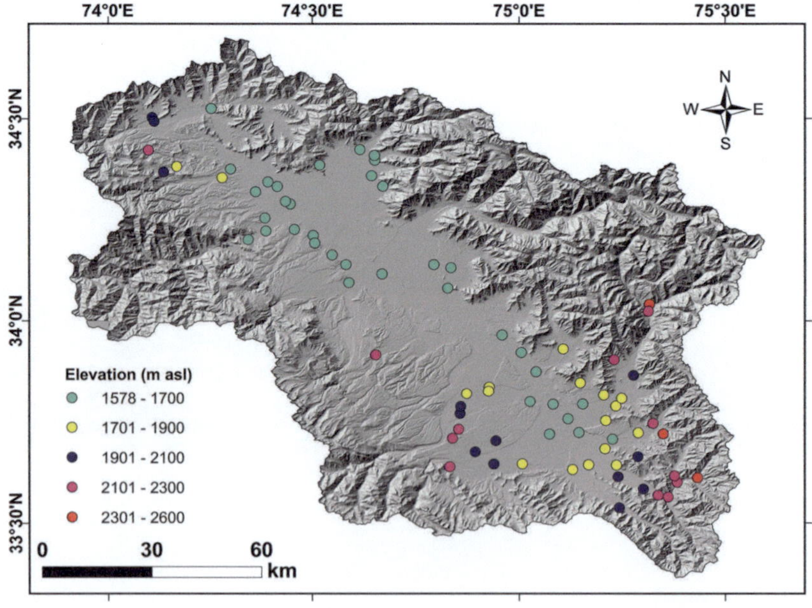

Cirsium arvense (**a**) Habit & Habitat, (**b**) Leaf, (**c**) Inflorescence,(**d**) Capitula, (**e**) Florets

Conium maculatum L., Sp. Pl. 1: 243 (1753).

Family	Apiaceae
English names	Poison hemlock, Devil's porridge
Local name	'Mori gassi'
Habit	Biennial herb, 200–300 cm tall
Stem	Erect, branched, grooved with irregular purple spots
Leaves	Alternate, petiolate, leaf-blade bi-tripinnate, ultimate segments ovate to elliptic, base sheathing, margin serrate
Inflorescence	Umbel
Flower	Small, sepals 5, calyx teeth obsolete, petals 5, white
Fruit	Mericarp
Pollination	Entomophily
Seed dispersal	Anemochory
Habitat	Agri-fields, forests, roadsides, gardens, orchards, riparian areas
Current status	Invasive
Impacts	Decreases native plant diversity, reduces quality and quantity of crop yield, causes economic loss in agriculture, highly toxic, reduces forage value of pastures
Nativity	Africa, Asia-Temperate
Global distribution	Africa, Asia-Temperate, Asia-Tropical, Europe, Northern America, Pacific, Southern America, Australasia

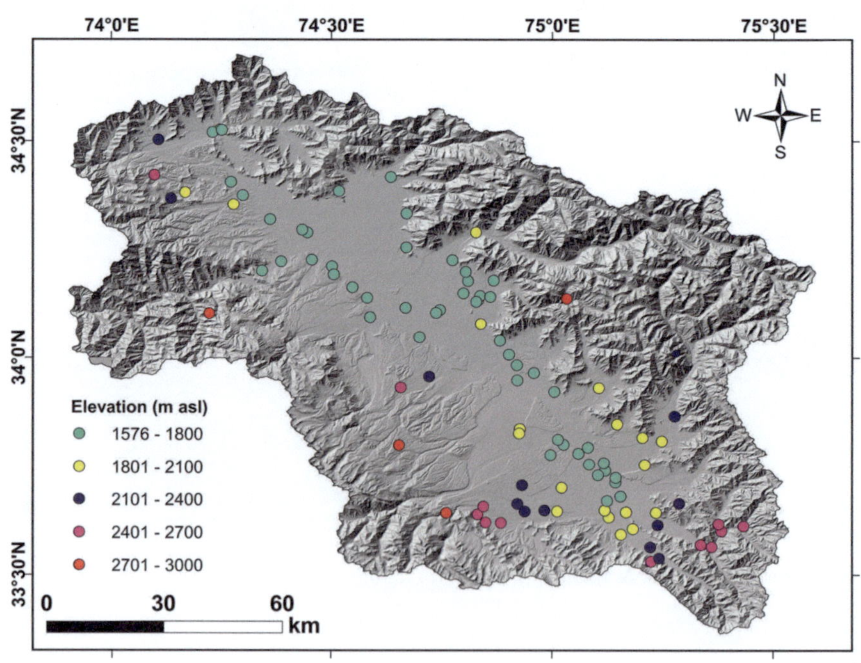

Conium maculatum: (**a**, **b**) Habit & Habitat, (**c**) Stem, (**d**) Leaf, (**e**) Inflorescence, (**f**) Fruit

Coreopsis tinctoria Nutt., J. Acad. Nat. Sci. Philadelphia 2: 114 (1821).

Family	Asteraceae
English name	Golden coreopsis
Local name	'Kari heafiz'
Habit	Annual herb, 10–150 cm tall
Stem	Erect, branched, glabrous
Leaves	Opposite, petiolate, leaf-blade 1–3 pinnate, lobes linear to lanceolate
Inflorescence	Capitula, involucre bell-shaped, phyllaries ovate to lanceolate
Flower	Ray florets yellow with red-brown blotch, disc florets red brown
Fruit	Achene
Pollination	Entomophily
Seed dispersal	Anemochory, ornithochory
Habitat	Gardens, roadsides, forests
Current status	Casual
Native range	Northern America
Global distribution	Africa, Asia-Temperate, Europe, Northern America, Southern America, Australasia

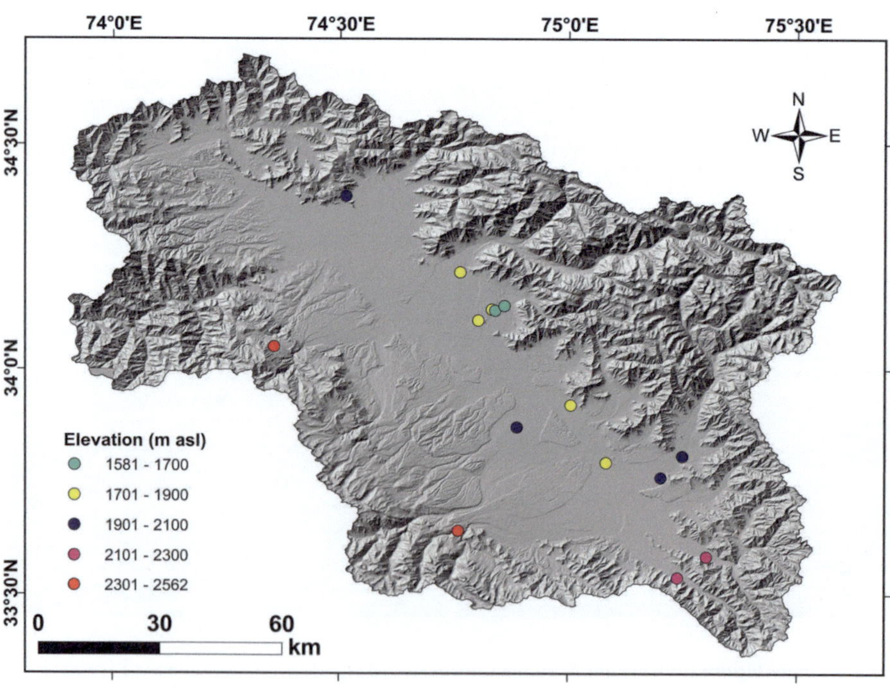

Coreopsis tinctoria: (**a**) Habit & Habitat, (**b**) Leaf, (**c, d**) Flower (Inflorescence)

Cosmos sulphureus Cav., Icon. [Cavanilles] 1(3): 56, t. 79 (1791).

Family	Asteraceae
English name	Sulphur cosmos
Local name	'Lyodur cosmas'
Habit	Annual herb, 100–300 cm tall
Stem	Erect, branched, pubescent
Leaves	Opposite, petiolate, pinnately divided, margin ciliate, apex mucronate
Inflorescence	Capitula, phyllaries oblong to lanceolate
Flower	Ray florets deeply yellow to intensely orange, disc florets tubular, corolla orange or yellow
Fruit	Achene
Pollination	Entomophily
Seed dispersal	Autochory, anemochory
Habitat	Grasslands, roadsides, gardens, orchards
Current status	Casual
Impacts	Decreases native plant diversity; reduces forage value of pastures
Native range	Northern America
Global distribution	Africa, Asia-Tropical, Northern America, Southern America

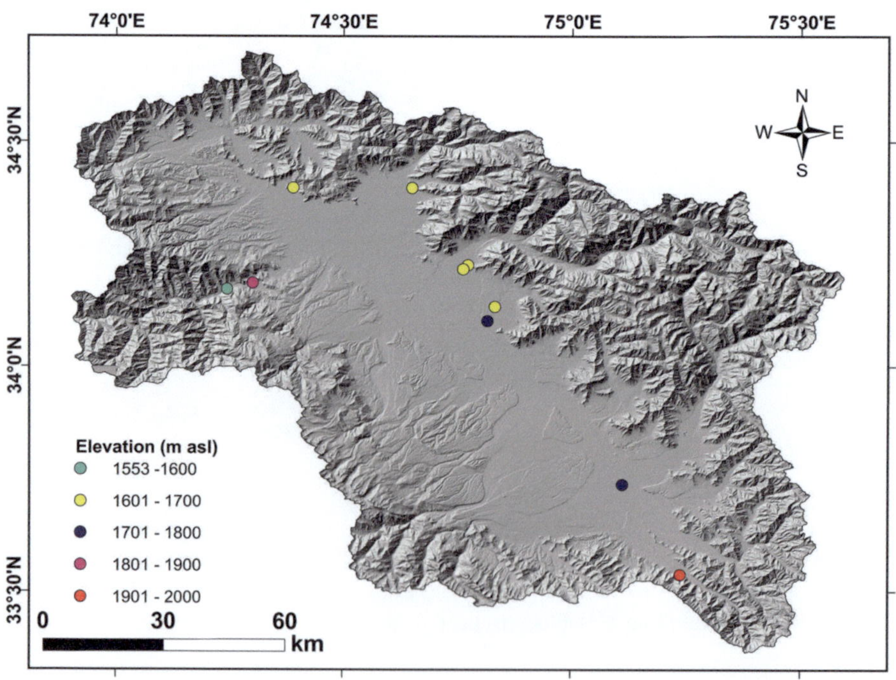

Cosmos sulphureus: (**a**) Habit, (**b**) Leaf, (**c**) Floral bud, (**d**, **e**) Flower (Inflorescence), (**f**, **g**) Fruit

Datura stramonium L., Sp. Pl. 1: 179 (1753).

Family	Solanaceae
English names	Jimson weed, Thorn apple
Local name	'Dati-er kul'
Habit	Annual herb, 60–150 cm tall
Stem	Erect, branched, glabrous
Leaves	Petiolate, leaf-blade large, ovate, margin irregularly undulate, dentate
Inflorescence	Solitary
Flower	Trumpet-shaped, fragrant, calyx long, tubular, corolla white or purple, shallowly five-lobed, each lobe with acuminate tip
Fruit	Capsule, spiny
Pollination	Entomophily
Seed dispersal	Ornithochory
Habitat	Agri-fields, roadsides, grasslands, gardens, forests, riparian areas
Current status	Invasive
Impacts	Reduces quality and quantity of crop yield, reported to act as an alternative host of pests and pathogens of many crops, poisonous to humans and livestock; decreases the native plant diversity
Native range	Northern America
Global distribution	Northern America, Southern America

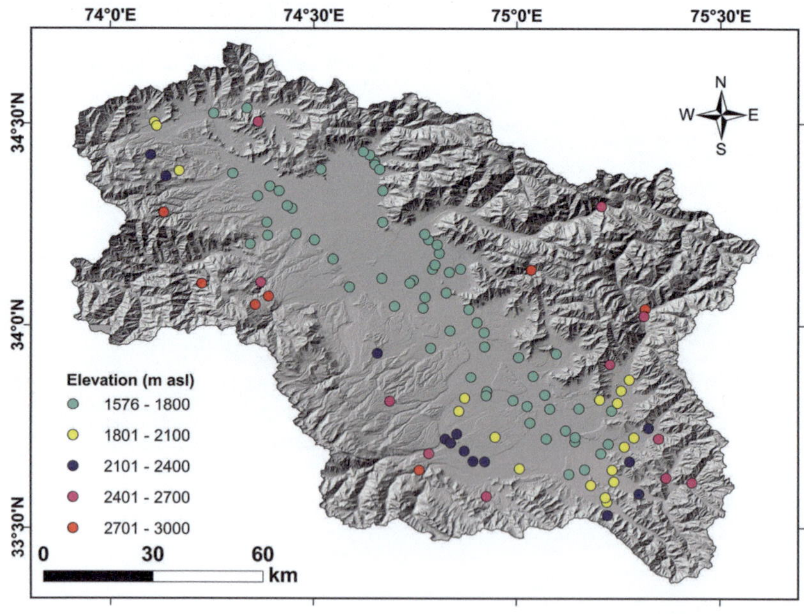

Datura stramonium: (**a**) Habit, (**b**) Leaf, (**c**) Flower, (**d**) Fruit

Dianthus deltoides L., Sp. Pl. 1: 411 (1753).

Family	Caryophyllaceae
English name	Maiden pink
Local name	'Tikooni dianthus'
Habit	Perennial rhizomatous herb
Stem	Stems ascending to erect upto 25 cm tall
Leaves	Basal leaves oblanceolate, sessile, cauline leaves in pairs, sessile, linear to lanceolate, margin entire, apex acute
Inflorescence	Solitary or clustered at the top of the stem
Flower	Pedicellate, pedicel upto 3 cm in length, calyx veined, glabrous or sparsely pubescent, lobes linear to lanceolate, petals 5, deep pink to purple or white in colour, a darker band forms a ring near the base of petals, upper surface dotted, bearded, tip serrate
Fruit	Capsule
Pollination	Entomophily
Seed dispersal	Autochory, anemochory
Habitat	Gardens, grasslands, forests
Current status	Naturalised
Native range	Asia-Temperate, Europe
Global distribution	Asia-Temperate, Northern America, Southern America, Australasia

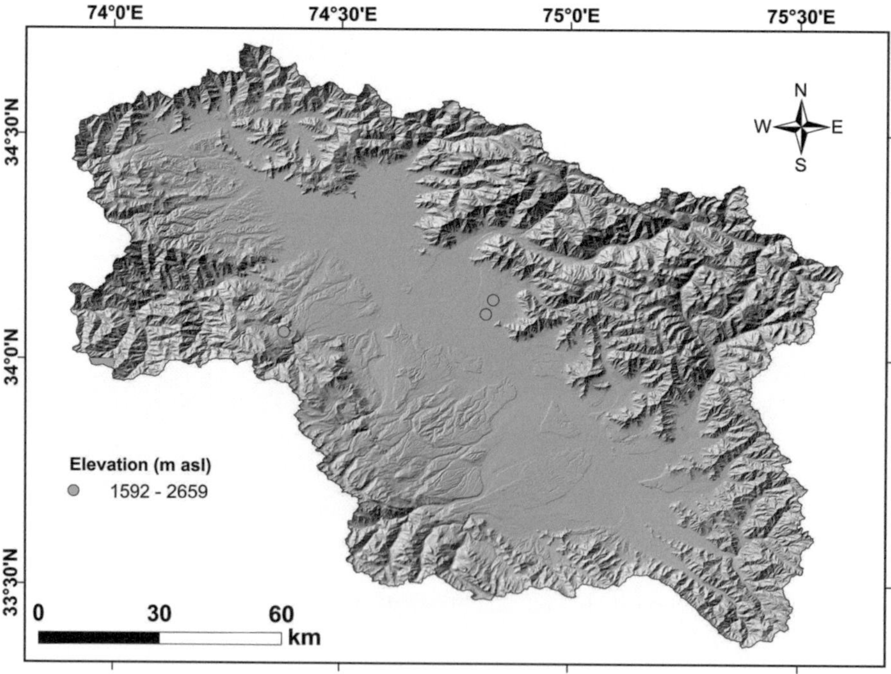

Dianthus deltoides: (**a**) Habit, (**b**) Rhizome, (**c**) Leaf, (**d**) Flower, (**e**) Fruit

Digitalis grandiflora Mill., Gard. Dict., ed. 8. [erratum page; Digitalis no. 4 in text] (1768).

Family	Plantaginaceae
English name	Large yellow foxglove
Local name	'Lyodur digitilis'
Habit	Perennial herb, upto 100 cm tall
Stem	Ascending to erect, pubescent
Leaves	Leaves ovate to lanceolate, sessile, pubescent, margin serrate, apex acute. Basal leaves in rosette and largest, upper stem leaves alternate and smaller in size
Inflorescence	Spike
Flower	Calyx green lobed, corolla creamy yellow, tubular, pendulous, pubescent with interior brown markings
Fruit	Capsule
Pollination	Entomophily
Seed dispersal	Anemochory
Habitat	Grasslands, gardens, forests, riparian areas
Current status	Naturalised
Impacts	Decreases native plant diversity, reduces the forage value of pastures
Nativity	Asia-Temperate, Europe
Global distribution	Asia-Temperate, Europe, Northern America

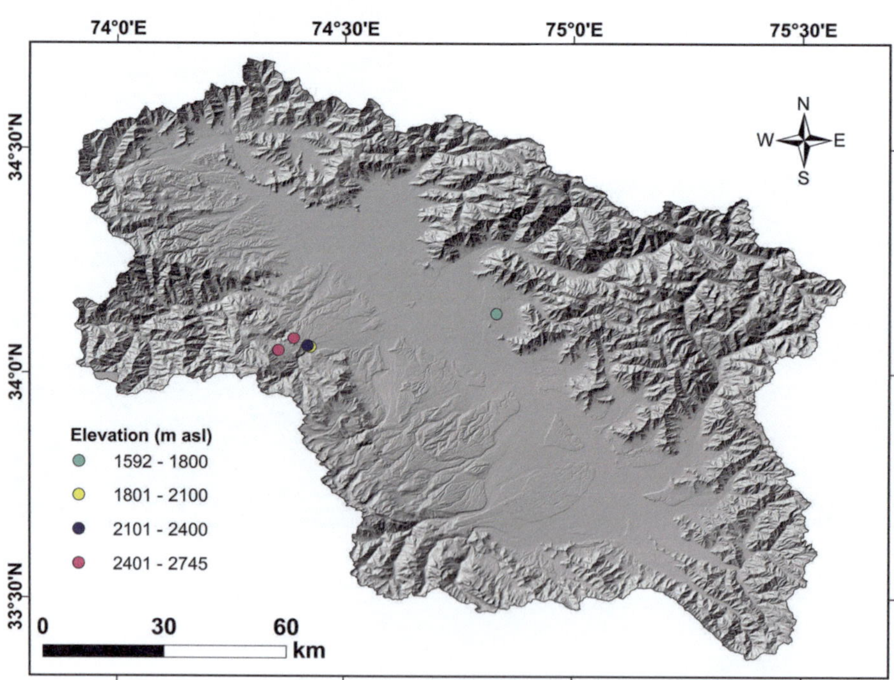

Digitalis grandiflora: (**a**) Habit, (**b**) Habitat, (**c**) Leaf, (**d**) Inflorescence (Flowers)

Digitalis purpurea L., Sp. pl. 2: 621 (1753).

Family	Plantaginaceae
English name	Common foxglove
Local name	'Aam digitilis'
Habit	Biennial or perennial herb, 60–170 cm tall
Stem	Erect, unbranched, pubescent
Leaves	Basal leaves rosulate, petiole narrowly winged, margin crenate, apex acuminate, stem leaves alternate, smaller and sessile
Inflorescence	Cyme, 10–20 flowered, rarely branched, flowers drooping
Flower	Calyx campanulate, five-lobed, corolla purple to white, inside spotted, lobe apex pubescent
Fruit	Capsule
Pollination	Entomophily
Seed dispersal	Anemochory
Habitat	Grasslands, gardens, forests, riparian areas
Current status	Naturalised
Impacts	Decreases native plant diversity; the leaves, flowers and seeds contain the cardiac glycoside digitoxin, which is extremely dangerous to humans and livestock and can be fatal if consumed
Native range	Africa, Europe
Global distribution	Africa, Asia-Tropical, Europe

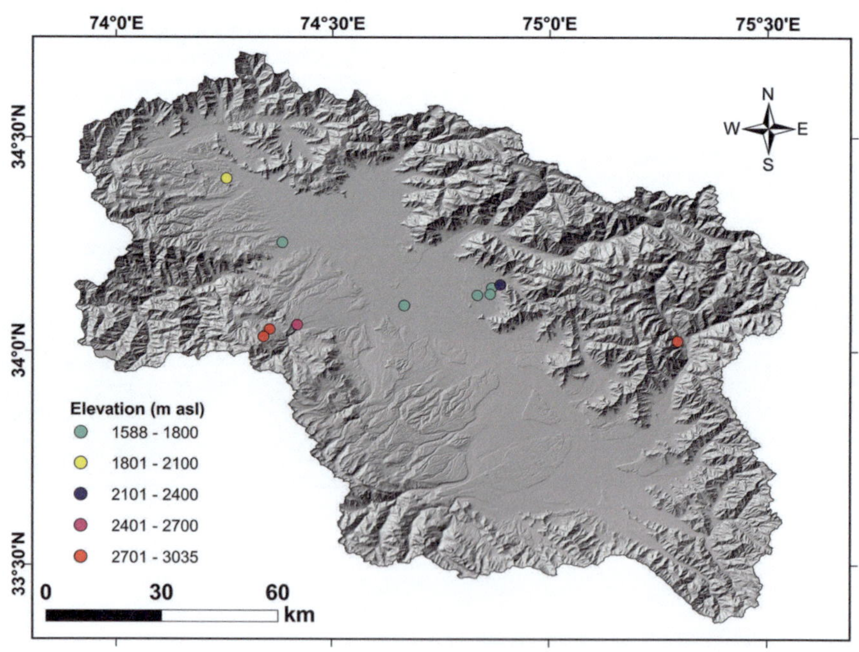

Digitalis purpurea: (**a**) Habit, (**b**) Leaf, (**c**) Floral bud, (**d**, **e**) Inflorescence, (**f**) Flower

Dysphania ambrosioides (L.) Mosyakin & Clemants, Ukrayins'k. Bot. Zhurn. 59 (4): 382 (2002).

Synonym	*Chenopodium ambrosioides* L.
Family	Amaranthaceae
English name	Mexican tea
Local name	'Zahear lyisi'
Habit	Annual or short-lived perennial herb, upto 120 cm tall
Stem	Erect, branched, hairy, aromatic
Leaves	Alternate, petiolate, leaf-blade lanceolate to elliptic, glands on lower surface, margin coarsely serrate or entire
Inflorescence	Panicle
Flower	3–5 flowers per glomerule, perianth greenish, segments 3–4 in female flowers, 4–5 in bisexual flowers, pubescent or glabrous, variably connate
Fruit	Utricle
Pollination	Anemophily
Seed dispersal	Anemochory, zoochory
Habitat	Agri-fields, roadsides, gardens, grasslands, orchards, forests, hillslopes
Current status	Invasive
Impacts	Reduces the yield of crops like maize, rice and pulses; reported to act as a host to many fungal diseases, allergic to humans; decreases native plant diversity
Native range	Northern America, Southern America
Global distribution	Africa, Asia-Temperate, Asia-Tropical, Northern America, Southern America

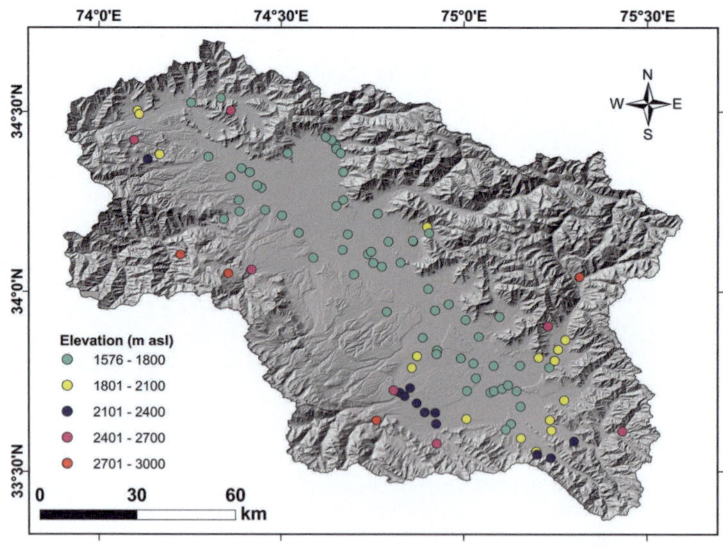

Dysphania ambrosioides: (**a**) Habitat, (**b**) Habit, (**c**) Leaf, (**d**) Inflorescence

Erigeron annuus (L.) Pers., Syn. Pl. [Persoon] 2 (2): 431 (1807).

Family	Asteraceae
English names	Annual fleabane, Daisy fleabane
Local name	'Dazy shaal-i lot'
Habit	Annual herb, upto 150 cm tall
Stem	Erect, hairy, simple or branched above
Leaves	Alternate, leaf-blade lanceolate, margin coarsely toothed to entire, hairy, lower stem leaves stalked, stalks winged, upper stem leaves stalkless
Inflorescence	Capitula born on leaf axils at the end of branches, more or less corymbose, involucre disc-shaped
Flower	Ray florets white or pinkish, disc florets yellow
Fruit	Cypsela
Pollination	Entomophily
Seed dispersal	Anemochory, zoochory
Habitat	Gardens, roadsides, grasslands, agri-fields, orchards, hillslopes
Current status	Invasive
Impacts	Decreases native plant diversity; reduces the yield of corn; allelopathic
Native range	Northern America
Global distribution	Africa, Asia-Temperate, Asia-Tropical, Europe, Northern America, Southern America, Australasia

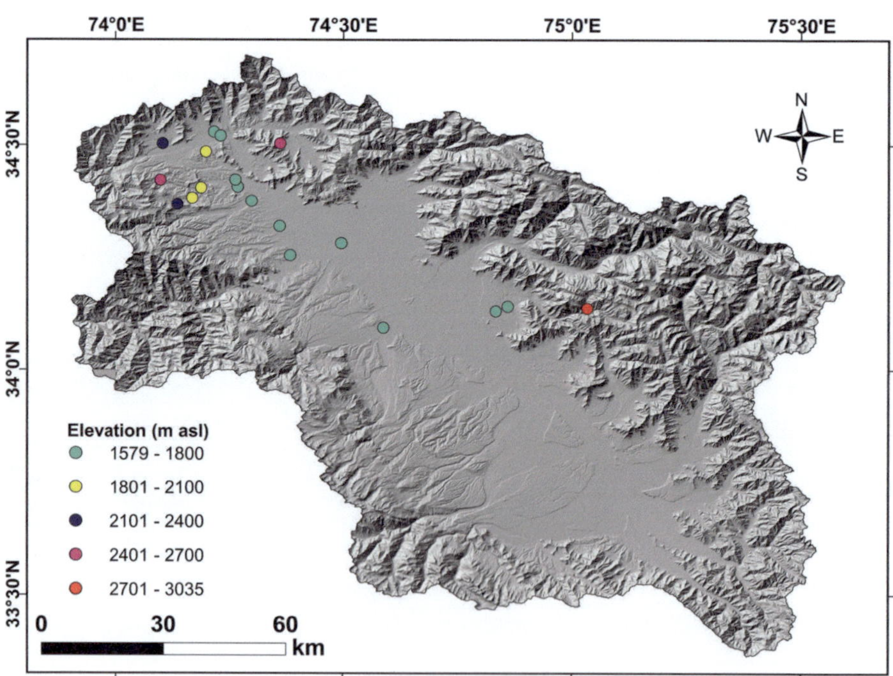

Erigeron annuus: (**a**) Habitat, (**b**) Habit, (**c**) Leaf, (**d**) Inflorescence (Flowers)

Erigeron canadensis L., Sp. Pl. 2: 863 (1753).

Family	Asteraceae
English names	Canadian horseweed, Canadian fleabane
Local name	'Shaal-i lot'
Habit	Annual herb, growing upto 150 cm tall
Stem	Erect, densely branched, sparsely hairy
Leaves	Basal leaves petiolate, leaf-blade oblanceolate, base attenuate, margin sparsely serrate or entire, apex acute or acuminate, mid and upper leaves sessile or subsessile, leaf-blade linear to lanceolate
Inflorescence	Capitula numerous, terminal, involucre subcylindrical
Flower	Ray florets white, disc florets yellowish
Fruit	Cypsela
Pollination	Entomophily
Seed dispersal	Anemochory, zoochory
Habitat	Grasslands, roadsides, gardens, orchards, agri-fields, forests, hillslopes
Current status	Invasive
Impacts	Decreases native plant diversity; reduces the quality and quantity of crop yield; reported to act as a host to many fungal and viral diseases; alters the ratio of carbon and nitrogen in soil
Native range	Northern America, Southern America
Global distribution	Africa, Asia-Temperate, Asia-Tropical, Europe, Northern America, Pacific, Southern America

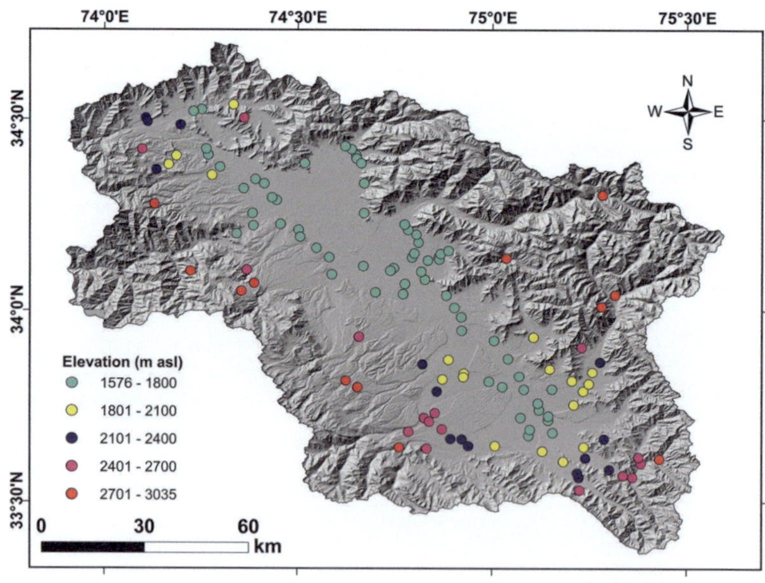

Erigeron canadensis: (**a**) Habitat, (**b**) Habit, (**c**) Leaf, (**d**) Inflorescence

Erigeron sumatrensis Retz., Observ. Bot. (Retzius) 5: 28 (1788).

Family	Asteraceae
English name	Tall fleabane
Local name	'Thod shaal-i lot'

Habit	Annual herb, upto 200 cm tall
Stem	Erect, branched at the top, leafy, hairy
Leaves	Basal leaves withered at anthesis, lower stem leaves petiolate, blade lanceolate to oblanceolate, surface strigose, base attenuate, margin serrate, upper stem leaves sessile, blade lanceolate to linear, margin serrate or entire
Inflorescence	Capitula in panicles, involucre campanulate or urn-shaped, phyllaries 3-seriate, hairy
Flower	Ray florets numerous, yellowish or creamy, disc florets 6–11, yellowish
Fruit	Cypsela
Pollination	Entomophily
Seed dispersal	Anemochory, zoochory
Habitat	Agri-fields, roadsides, gardens, grasslands, orchards, forests, hillslopes

Current status	Invasive
Impacts	Reduces the quality and quantity of crop yield; reported to act as a host to many pathogens, decreases native plant diversity

Nativity	Southern America
Global distribution	Africa, Asia-Temperate, Asia-Tropical, Europe, Southern America, Australasia

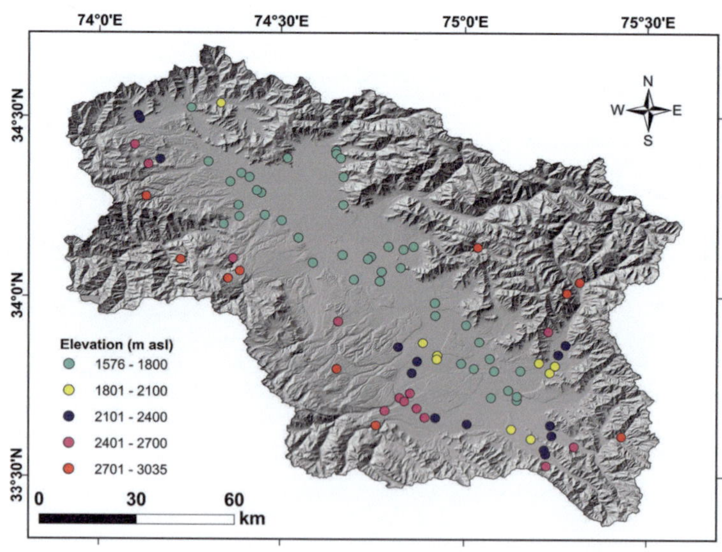

Erigeron sumatrensis: (**a**) Habitat, (**b**) Leaf, (**c**) Inflorescence

Eryngium billardierei F. Delaroche, Nouv. Bull. Sci. Soc. Philom. Paris 1: 87 (1807).

Family	Apiaceae
English name	Billardiere's eryngo
Local name	'Kiend gazri-gassi'
Habit	Perennial herb, 50–80 cm tall
Stem	Erect, branched, bluish
Leaves	Basal leaves petiolate, stem leaves sessile, leaf-blade pinnately divided, margin spiny
Inflorescence	Umbel, involucre of 6–8 bracts alternating with spines
Flower	Sessile, sepals lanceolate to ovate, greenish, petals violet blue
Fruit	Schizocarp
Pollination	Entomophily
Seed dispersal	Autochory, zoochory
Habitat	Grasslands, roadsides, orchards, agri-fields, hillslopes
Current status	Invasive
Impacts	Declines forage value of pastures, decreases the native plant diversity; prickles damage clothes and skin of humans
Native range	Asia-Temperate, Asia-Tropical
Global distribution	Asia-Temperate, Asia-Tropical, Europe

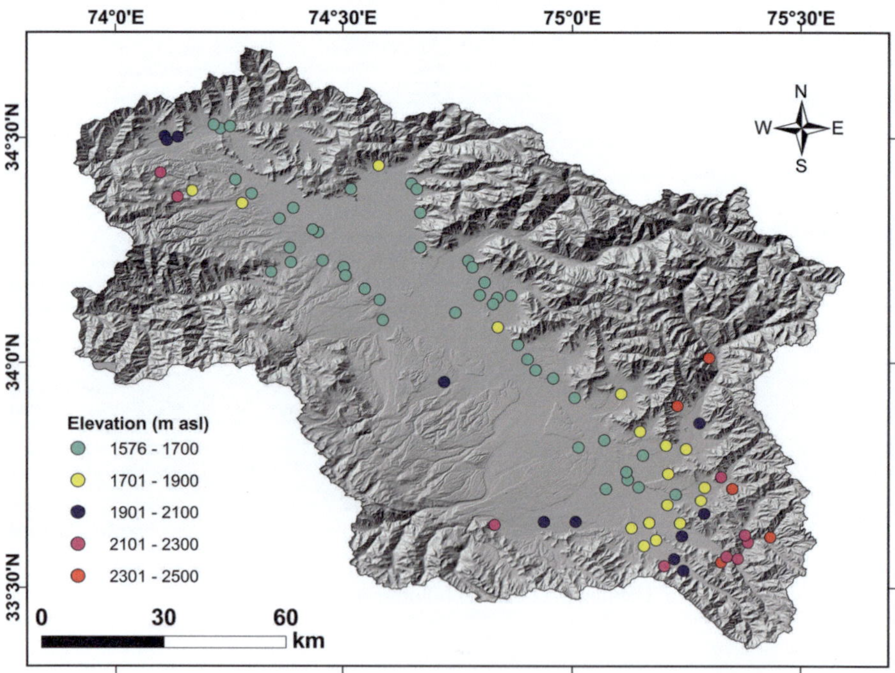

Eryngium billardierei: (**a**) Habit, (**b**) Inflorescence, (**c**) Flower

Euphorbia esula L., Sp. pl. 1: 461 (1753).

Family	Euphorbiaceae
English name	Leafy spurge
Local name	'Vather goer-gassi'
Habit	Perennial herb, 100–150 cm tall
Stem	Single or many, branched from base, smooth
Leaves	Alternate, sessile, leaf-blade linear to lanceolate, margin slightly wavy
Inflorescence	Cyathium in pseudo-umbel, involucre consists of leaf-like bracts
Flower	One female flower surrounded by many male flowers, ovary cup-shaped
Fruit	Capsule
Pollination	Entomophily
Seed dispersal	Autochory, zoochory
Habitat	Forests, roadsides, grasslands, gardens, orchards
Current status	Naturalised
Impacts	Decreases the native plant diversity; declines the forage value of pastures; reduces recreational value of tourist spots and natural areas; latex produced is annoying to common public and tourists
Native range	Africa, Asia-Temperate, Asia-Tropical, Europe
Global distribution	Africa, Asia-Temperate, Asia-Tropical, Europe, Northern America

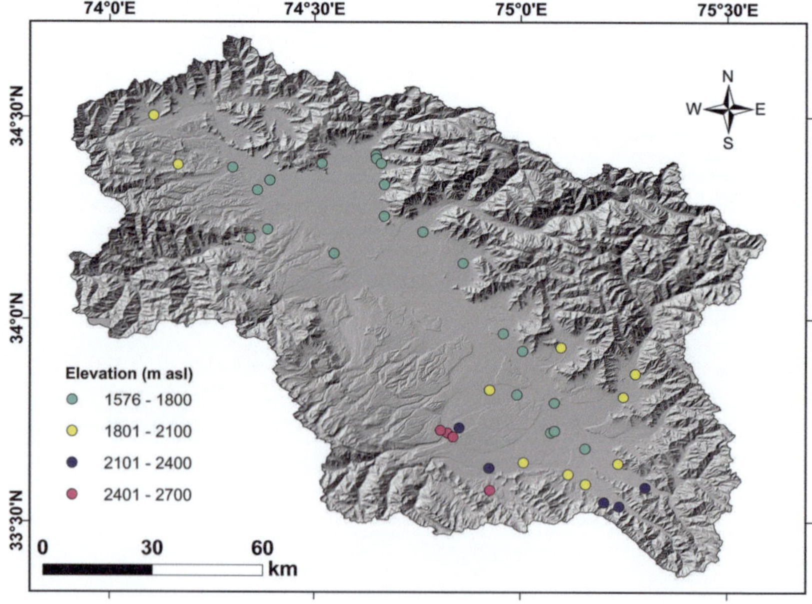

Euphorbia esula: (**a**) Habitat, (**b**) Habit, (**c**) Inflorescence, (**d**) Fruit

Euphorbia lathyris L., Sp. Pl. 1: 457 (1753).

Family	Euphorbiaceae
English name	Caper spurge
Local name	'Panjeab goer-gassi'
Habit	Annual or biennial herb, 100–150 cm tall.
Stem	Erect, branched, bluish green, glabrous
Leaves	Opposite, decussate, sessile, leaf-blade lanceolate, glabrous, base clasping, apex acute or acuminate, margin entire
Inflorescence	Pseudo-umbel, cyathium subsessile, involucre sub-campanulate
Flower	Green or yellowish green, petals absent
Fruit	Capsule
Pollination	Entomophily
Seed dispersal	Anemochory, zoochory
Habitat	Gardens, roadsides, grasslands, riparian areas
Current status	Naturalised
Impacts	Decreases the native plant diversity; reduces forage value of pastures, poisonous to humans and livestock; reported to cause dermatitis and eye irritation
Native range	Asia-Temperate, Europe
Global distribution	Asia-Temperate, Asia-Tropical, Europe, Northern America, Southern America, Australasia

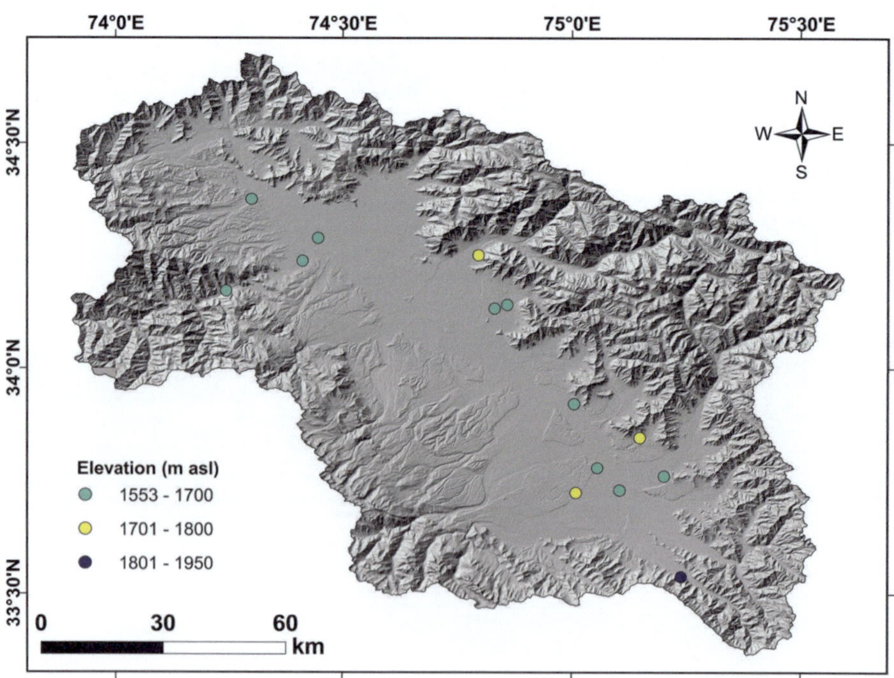

Euphorbia lathyris: (**a**) Habit & Habitat, (**b**) Leaf, (**c**) Inflorescence, (**d**) Fruit

Euphorbia prostrata Aiton., Hort. Kew. 2: 139 (1789).

Family	Euphorbiaceae
English name	Prostrate sandmat
Local name	'Patheir goer-gassi'
Habit	Annual herb, upto 20 cm tall
Stem	Many, prostrate, light red, glabrous or sparsely hairy
Leaves	Opposite, sessile or subsessile, leaf-blade oval, margin finely serrate, stipules connate
Inflorescence	Cyathium, involucre turbinate, hairy
Flower	One female flower, pedicellate, stigma two-lobed, many short male flowers
Fruit	Capsule
Pollination	Entomophily
Seed dispersal	Autochory, hydrochory
Habitat	Grasslands, agri-fields, roadsides, gardens
Current status	Naturalised
Impacts	Decreases the native plant diversity; mostly found as a weed of maize crop and reduces its yield
Native range	Northern America, Southern America
Global distribution	Africa, Asia-Temperate, Asia-Tropical, Europe, Northern America, Pacific, Southern America, Australasia

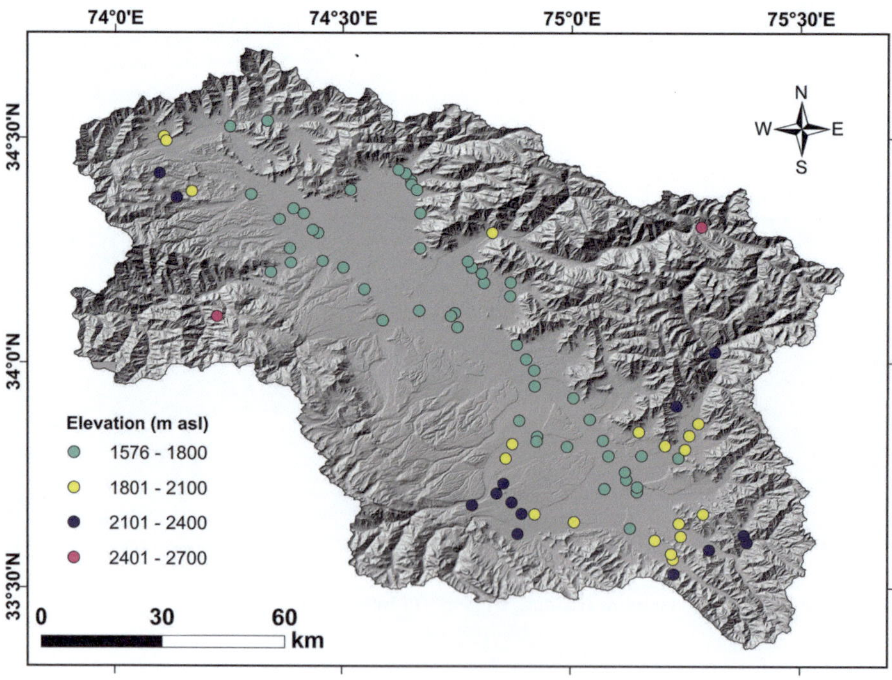

Euphorbia prostrata: (**a**) Habit & Habitat, (**b**) Inflorescence, (**c**) Leaf, (**d**) Flower

Galinsoga parviflora Cav., Icon. 3(2): 41–42, pl. 281. (1794).

Family	Asteraceae
English names	Quick weed, Gallant soldier
Local name	'Naram nod'
Habit	Annual herb, 30–100 cm tall
Stem	Erect, branched, glabrous to sparsely pubescent
Leaves	Opposite, petiolate, leaf-blade ovate, margin coarsely serrate
Inflorescence	Capitula, peduncles 1–3.5 cm long, involucre campanulate
Flower	Ray florets 4–6 per head, white, three-lobed, disc florets yellow
Fruit	Achene
Pollination	Entomophily
Seed dispersal	Anemochory, zoochory
Habitat	Agri-fields, roadsides, grasslands, orchards, hill slopes, forests
Current status	Invasive
Impacts	Decreases the native plant diversity; allelopathic, reduce the agricultural productivity, especially the maize and vegetable crops
Native range	Northern America, Southern America
Global distribution	Africa, Asia-Temperate, Asia-Tropical, Europe, Northern America, Southern America, Pacific, Australasia

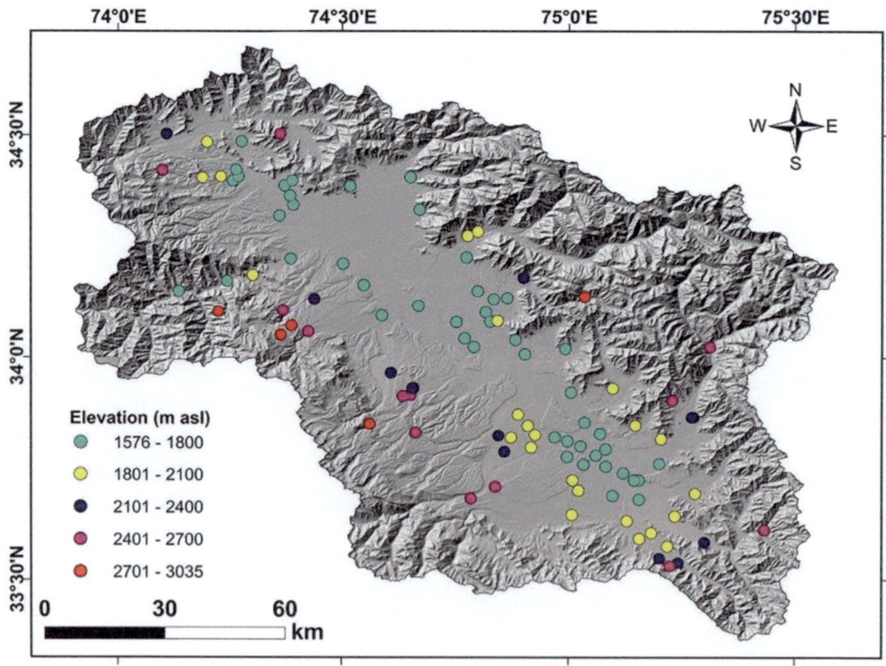

Galinsoga parviflora: (**a**) Habitat, (**b**) Habit, (**c, d**) Flower (Inflorescence)

Helianthus tuberosus L., Sp. Pl. 2: 905 (1753).

Family	Asteraceae
English name	Jerusalem artichoke
Local name	'Farm gogji'
Habit	Perennial herb, 200–250 cm tall
Stem	Erect, scabrid to hirsute
Leaves	Lower leaves opposite, upper alternate, petiolate, leaf-blade ovate, rough or hairy, margin entire or serrate
Inflorescence	Capitula, involucre hemispherical
Flower	Ray floret lamina oblanceolate, yellow, disc florets light yellow
Fruit	Achene
Pollination	Entomophily
Seed dispersal	Ornithochory, zoochory
Habitat	Gardens, roadsides, agri-fields, riparian areas
Current status	Naturalised
Impacts	Decreases the native plant diversity; allelopathic, declines the agricultural productivity
Native range	Northern America
Global distribution	Africa, Asia-Temperate, Europe, Northern America, Southern America, Australasia

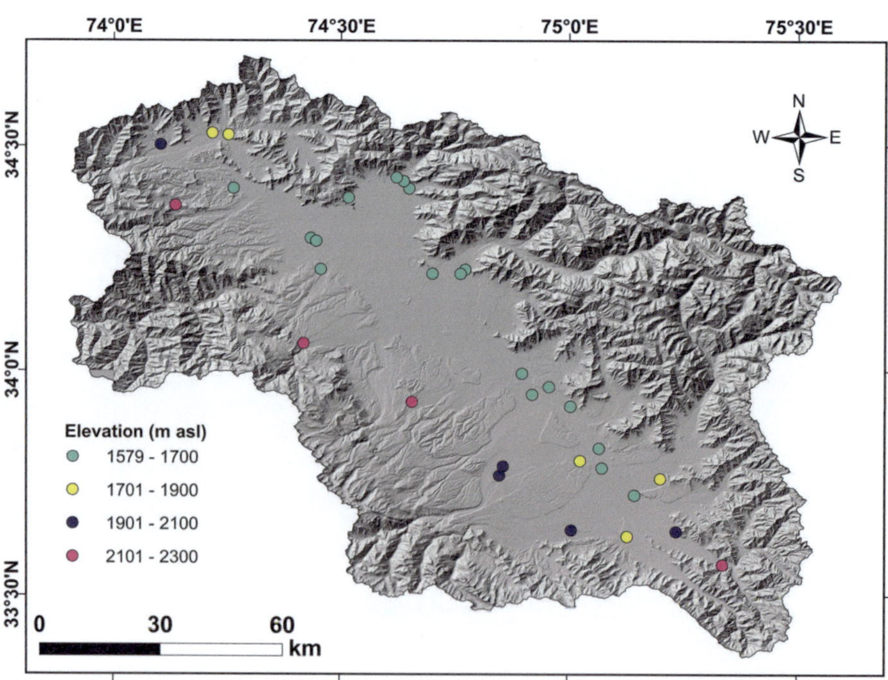

Helianthus tuberosus: (**a**) Habit & Habitat, (**b**) Stem, (**c**) Leaf, (**d, e**) Flower (Inflorescence)

Hemerocallis fulva L., Sp. Pl., ed. 2. 1: 462 (1762).

Family	Asphodelaceae
English names	Orange daylily, Tawny daylily
Local name	'Sangtari daylily'
Habit	Perennial herb, 40–180 cm tall
Stem	Flowering stems erect, hollow, sterile bracts present
Leaves	Large, leaf-blade linear, apex acute, margin entire
Inflorescence	Cyme
Flower	Pedicellate, bisexual, showy, perianth tube short, stout, tepals orange, stamens petaloid
Fruit	Capsule
Pollination	Entomophily
Seed dispersal	Autochory
Habitat	Forests, gardens, roadsides, orchards, riparian areas
Current status	Naturalised
Impacts	Decreases the native plant diversity
Native range	Asia-Temperate
Global distribution	Asia-Temperate, Asia-Tropical, Northern America, Australasia

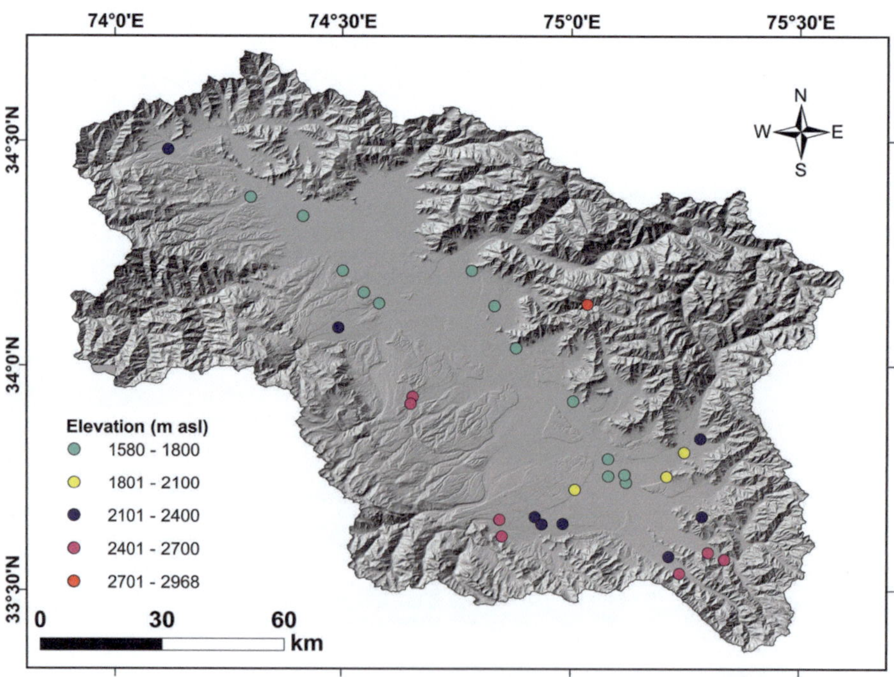

Hemerocallis fulva: (**a**) Habit & Habitat, (**b**) Leaf, (**c**) Floral bud, (**d**) Flower

Hesperis matronalis L., Sp. Pl. 2: 663 (1753).

Family	Brassicaceae
English name	Dame's rocket
Local name	'Wangien rocket'
Habit	Biennial or perennial herb, 40–120 cm tall
Stem	Erect, branched, leafy, hairy
Leaves	Alternate, sessile or subsessile, leaf-blade lanceolate, base amplexicaul, margin dentate
Inflorescence	Raceme
Flower	Calyx tubular, sepals oblong, corolla pinkish, lavender or violet, petals 4, stamens 6, stigma two-lobed
Fruit	Siliqua
Pollination	Entomophily
Seed dispersal	Ornithochory, zoochory
Habitat	Grasslands, roadsides, gardens, forests
Current status	Naturalised
Impacts	Decreases the native plant diversity; reduces the quantity and quality of crop yield
Native range	Asia-Temperate, Europe
Global distribution	Asia-Temperate, Europe, Northern America, Southern America, Australasia

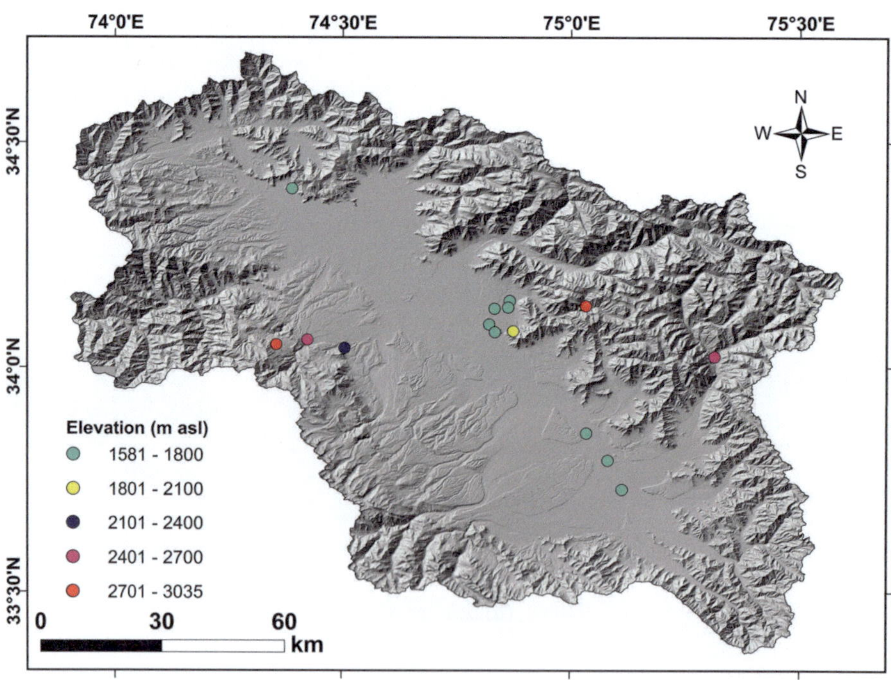

***Hesperis matronalis*:** (**a**) Habit, (**b**) Leaf, (**c**) Inflorescence, (**d**) Flower, (**e**) Fruit

Hibiscus trionum L., Sp. Pl. 2: 679 (1753).

Family	Malvaceae
English names	Flower-of-an-hour, Bladder hibiscus
Local name	'Foki-dooen hibiscus'
Habit	Annual herb, upto 100 cm tall
Stem	Erect, sparsely setose, branches prostrate
Leaves	Alternate, petiolate, palmatifid, ovate, glabrous or sparsely hairy, margin dentate, stipules linear, glabrous or sparsely hairy
Inflorescence	Solitary, axillary
Flower	Epicalyx of 7–12 segments, calyx twice the length of epicalyx, corolla white to pale yellow with a purplish base.
Fruit	Capsule
Pollination	Entomophily, autogamy
Seed dispersal	Anemochory, zoochory
Habitat	Agri-fields, roadsides, gardens, orchards
Current status	Naturalised
Impacts	Decreases native plant diversity; reduces the agricultural productivity
Native range	Africa, Asia-Temperate, Asia-Tropical, Europe
Global distribution	Africa, Asia-Temperate, Asia-Tropical, Europe

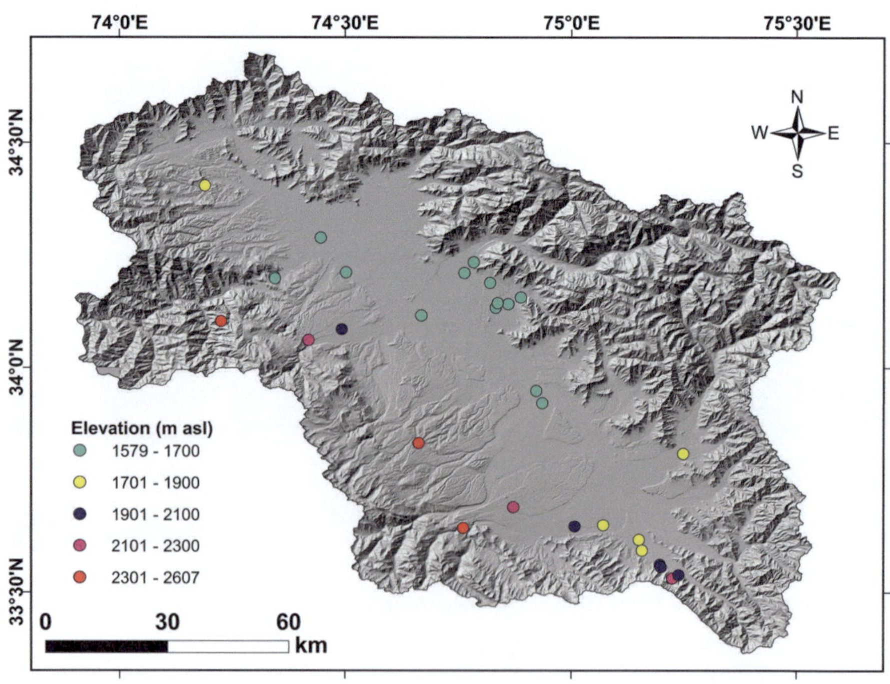

Hibiscus trionum: (**a**) Habit & Habitat, (**b**) Leaf, (**c**) Floral bud, (**d**) Flower, (**e**) Fruit

Ipomoea purpurea L., Roth., Bot. Abh. Beobacht. 27 (1787).

Family	Convolvulaceae
English name	Common morning glory
Local name	'Eshkipyiachaan'
Habit	Annual herb
Stem	Twining, weak, hairy
Leaves	Alternate, petiolate, leaf-blade cordate, pubescent on both the surfaces, apex acuminate, margin entire
Inflorescence	Solitary axillary or 2–5 flowered cymes
Flower	Sepals 5, free, corolla funnel-shaped, bluish to rosy-pink
Fruit	Capsule
Pollination	Entomophily, autogamy
Seed dispersal	Anemochory, ornithochory, hydrochory
Habitat	Orchards, roadsides, forests, gardens
Current status	Naturalised
Impacts	Decreases native plant diversity; reduces the agricultural productivity, seeds are reported to be highly poisonous
Native range	Tropical and sub-tropical America
Global distribution	Asia-Temperate, Asia-Tropical, Europe, Northern America, Southern America, Australasia

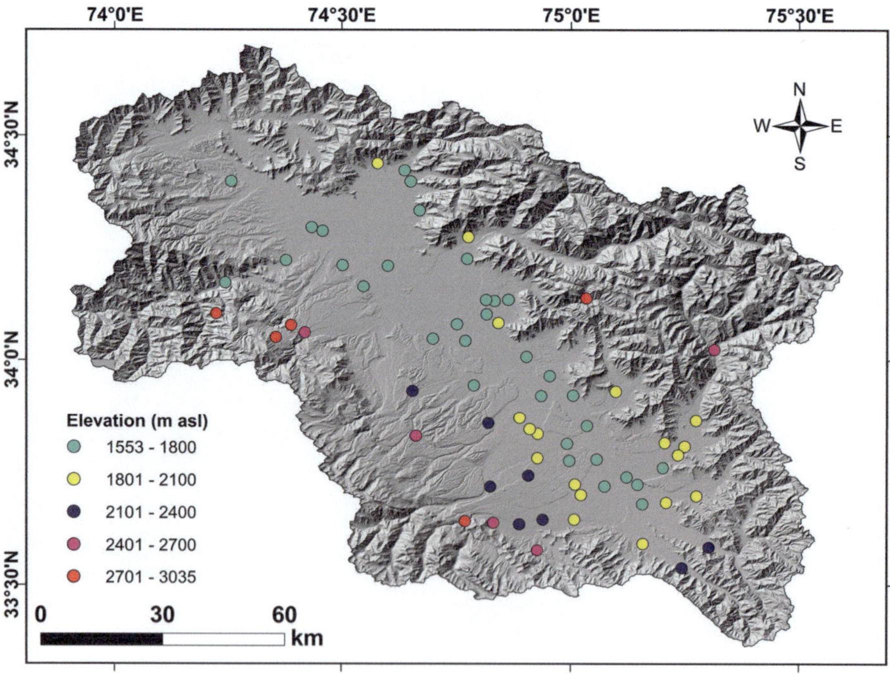

Ipomoea purpurea: (**a**) Habit & Habitat, (**b**) Leaf, (**c**) Floral bud, (**d, e**) Flower

Iris germanica L., Sp. Pl. 1: 38 (1753).

Family	Iridaceae
English names	Bearded iris, German iris
Local name	'Wangien mazar posh'
Habit	Perennial rhizomatous herb, 60–120 cm tall
Stem	Flowering stem upto 120 cm tall, branched, glabrous
Leaves	Basal upto 40 cm long, leaf-blade linear, sharp, glaucous
Inflorescence	3–6 flowers per peduncle
Flower	Perianth bluish or violet, obovate, base cuneate, beard white or pale blue, yellow tipped
Fruit	Capsule
Pollination	Entomophily
Seed dispersal	Anemochory, zoochory
Habitat	Gardens, roadsides, agri-fields, forests
Current status	Naturalised
Native range	Europe
Global distribution	Asia-Temperate, Asia-Tropical, Europe, Australasia

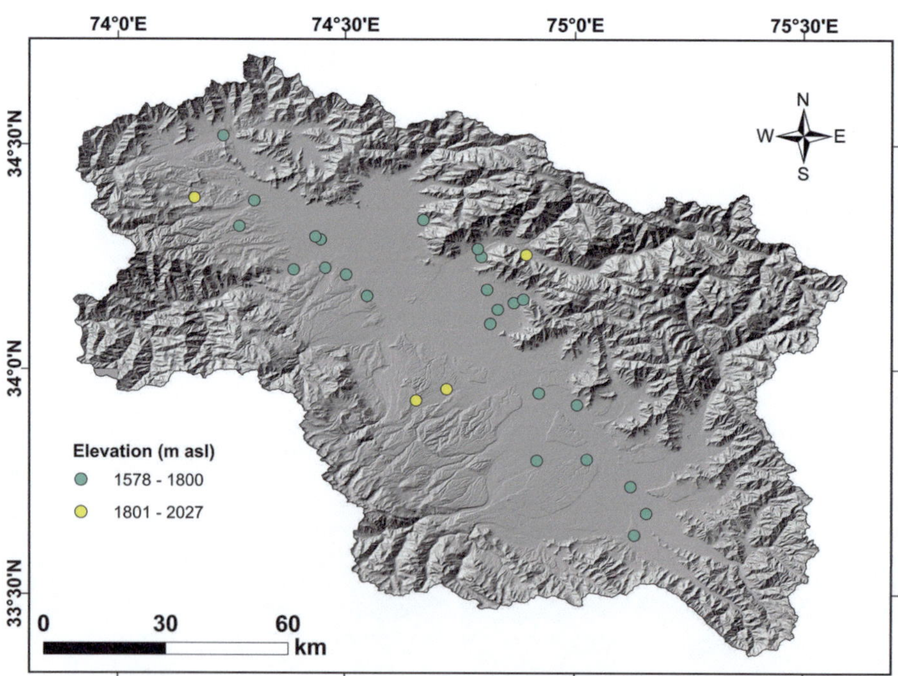

Iris germanica: (**a**) Habit & Habitat, (**b**) Flower, (**c**) Fruit

Iris pseudacorus L., Sp. Pl. 1: 38 (1753).

Family	Iridaceae
English names	Yellow iris, Yellow flag
Local name	'Lyedir mazar posh'
Habit	Perennial, rhizomatous herb, upto 180 cm tall
Stem	Erect, branched, glabrous
Leaves	Basal leaves deciduous, leaf-blade dark green sword-shaped with prominent thickening in the middle, upto 150 cm in length
Inflorescence	3–4 flowered, spathes green with brown margins, outer spathe strongly keeled, inner without keel
Flower	Sepals, bright yellow, sub-orbiculate, drooping, upper surface covered by brown lines, length upto 7.5 cm, petals without veins, lanceolate, style keeled, crests spreading, stigmas rounded with a prominent tongue
Fruit	Capsule
Pollination	Entomophily
Seed dispersal	Hydrochory
Habitat	Riparian areas
Current status	Casual
Impacts	Alters hydrology by reducing water flow and trapping sediments in its rhizomes; decreases the native plant diversity; reported to be highly poisonous to livestock, causes skin irritation in humans
Native range	Africa, Asia-Temperate, Europe
Global distribution	Africa, Asia-Temperate, Europe, Northern America, Southern America, Australasia

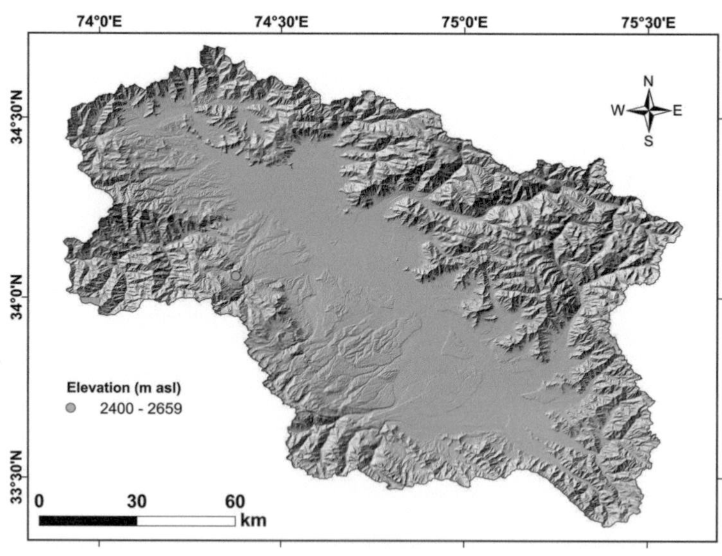

Iris pseudacorus: (**a**) Habitat, (**b**) Leaf, (**c**) Flower, (**d**) Spathe

Iris reticulata M.Bieb., Fl. Taur.-Cauc. 1: 34 (1808).

Family	Iridaceae
English names	Dwarf iris, Netted iris
Local name	'Zaeal mazar posh'
Habit	Perennial herb, bulb tunicated
Stem	Flowering stem 8–15 cm tall
Leaves	Arise from bulb, 2–4 per bulb, narrow with pointed apex
Inflorescence	Solitary
Flower	Perianth variable in colour—blue, purple or violet, covered with spathe, standards long, oblanceolate
Fruit	Capsule
Pollination	Entomophily
Seed dispersal	Anemochory, zoochory
Habitat	Agri-fields, roadsides, gardens, orchards, hill slopes
Current status	Naturalised
Impacts	Decreases native plant diversity; reduces the agricultural productivity
Native range	Asia-Temperate
Global distribution	Asia-Temperate

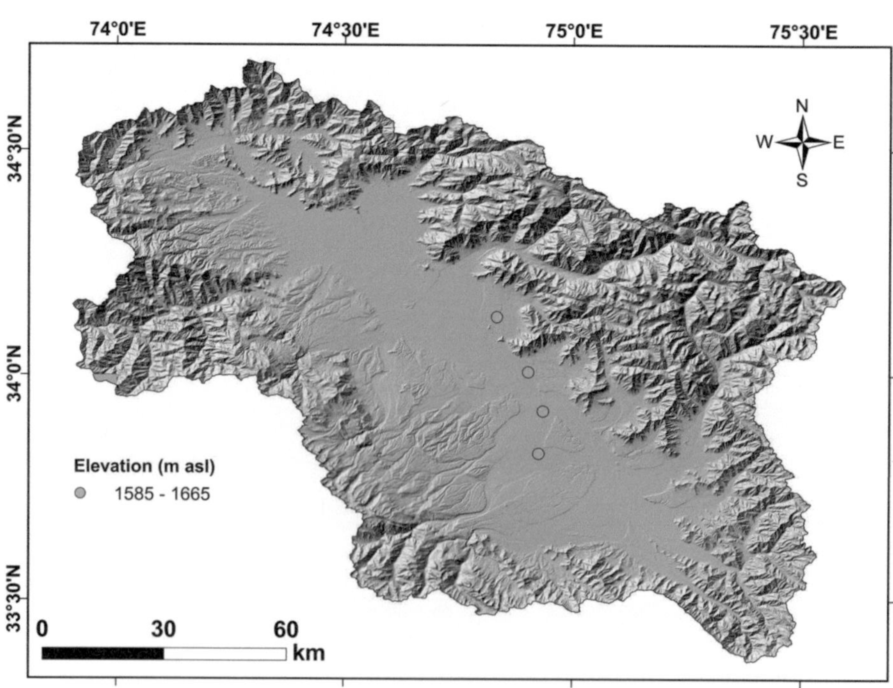

Iris reticulata: (**a**) Habitat, (**b**) Habit, (**c**) Flower, (**d**) Bulb

Lepidium virginicum L., Sp. P1. 2: 645 (1753).

Family	Brassicaceae
English name	Virginia pepper-weed
Local name	'Panjeab martchi gassi'

Habit	Annual or biennial herb, 10–70 cm tall
Stem	Erect, branched, hairy
Leaves	Basal leaves in rosette, petiolate, oblanceolate to spatulate, middle leaves linear to lanceolate, margin toothed, hairy, upper leaves linear
Inflorescence	Raceme
Flower	Sepals green, petals white, longer than sepals
Fruit	Silicula
Pollination	Entomophily
Seed dispersal	Anemochory
Habitat	Grasslands, roadsides, gardens, orchards, agri-fields

Current status	Invasive
Impacts	Decreases native plant diversity; reduces the agricultural productivity

Native range	Northern America, Southern America
Global distribution	Africa, Asia-Temperate, Asia-Tropical, Europe, Northern America, Pacific, Southern America, Australasia

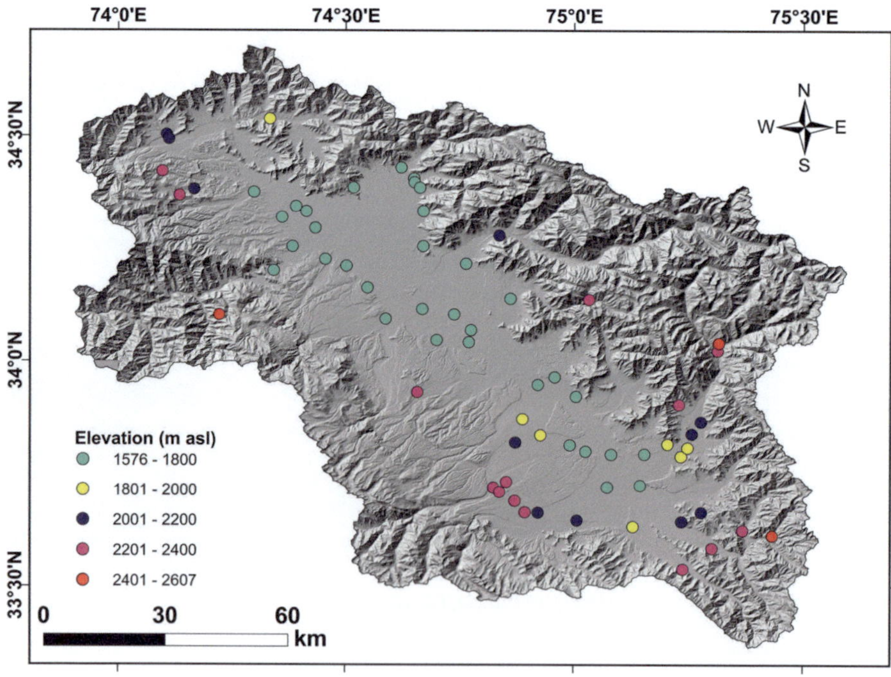

Lepidium virginicum: (**a**) Habit & Habitat, (**b**) Leaf, (**c**) Inflorescence, (**d**) Fruit

Leucanthemum vulgare Lam., Fl. Franc. (Lamarck) 2: 137 (1779).

Family	Asteraceae
English name	Oxeye Daisy
Local name	'Thod dazy'
Habit	Perennial herb, 15–100 cm tall
Stem	Erect, usually unbranched, glabrous or sparsely pubescent
Leaves	Basal leaves petiolate, leaf-blade spatulate or oblanceolate, base attenuate, margin crenulate, stem leaves alternate, sessile, leaf-blade obovate to lanceolate, base attenuate, margin dentate
Inflorescence	Capitula, involucre saucer-shaped, phyllaries ovate to lanceolate
Flower	Ray florets white and female, disc florets yellow and bisexual
Fruit	Achene
Pollination	Entomophily, autogamy
Seed dispersal	Anemochory
Habitat	Agri-fields, roadsides, gardens, grasslands, forests
Current status	Invasive
Impacts	Decreases native plant diversity; reduces the agricultural productivity, declines the forage value of pastures
Native range	Asia-Temperate, Europe
Global distribution	Africa, Asia-Temperate, Asia-Tropical, Europe, Northern America, Southern America, Australasia

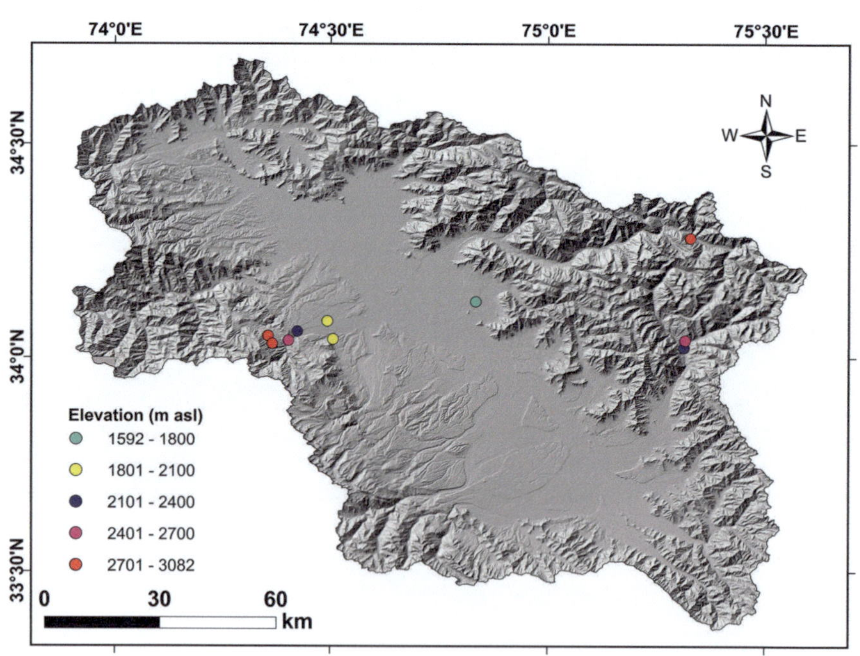

Leucanthemum vulgare: (**a**) Habit, (**b**) Basal leaf, (**c**) Cauline leaf, (**d**) Floral bud, (**e**) Flower (Inflorescence)

Linaria dalmatica (L.) Mill., Gard. Dict. (ed. 8) Linaria no. 13 (1768).

Family	Plantaginaceae
English name	Dalmatian toadflax
Local name	'Lyodur sheerdan'

Habit	Perennial herb, upto 120 cm tall
Stem	Erect, branched, glabrous
Leaves	Alternate, sessile, leaf-blade ovate, margin entire, base clasping
Inflorescence	Raceme
Flower	Dog flower like, spurred, calyx lobes linear to triangular, corolla bright yellow, lower lip hairy
Fruit	Capsule
Pollination	Entomophily
Seed dispersal	Anemochory, zoochory
Habitat	Grasslands, roadsides, gardens, forests, agri-fields, hillslopes

Current status	Naturalised
Impacts	Reduces the agricultural productivity, allelopathic, decreases native plant diversity; potentially toxic to livestock, reported to have a tendency to increase soil erosion

Native range	Asia-Temperate, Europe
Global distribution	Asia-Temperate, Asia-Tropical, Europe

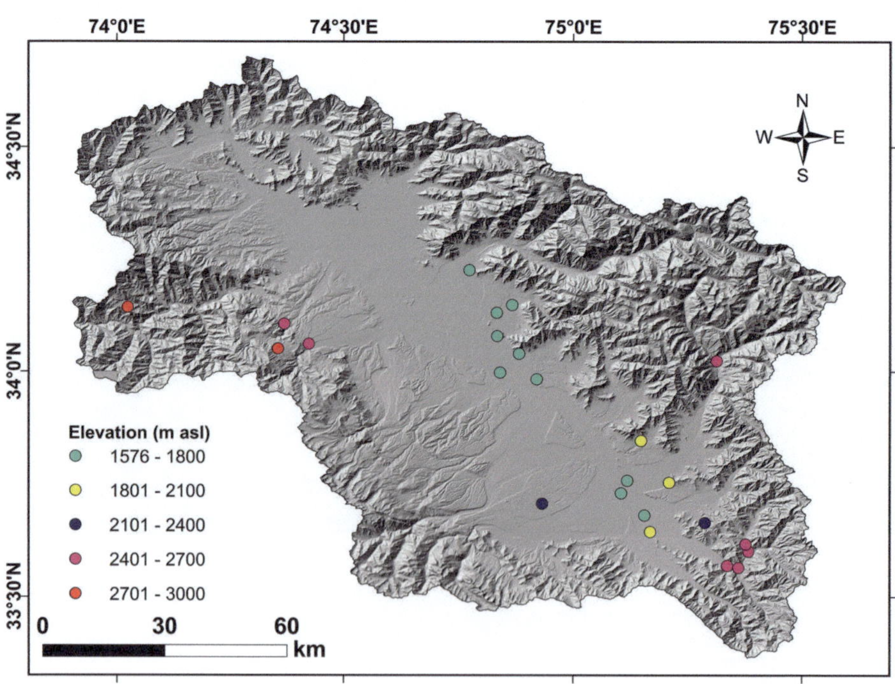

Linaria dalmatica: (**a**) Habit & Habitat, (**b**) Leaf, (**c**) Flower

Lupinus polyphyllus Lindl., Bot. Reg. 13: t. 1096 (1827).

Family	Fabaceae
English name	Garden lupin
Local name	'Bagi lupin'
Habit	Perennial herb, 30–150 cm tall
Stem	Erect, glabrous
Leaves	Alternate, petiolate, palmately compound, leaflets lanceolate, pubescent, margin entire, stipules adnate
Inflorescence	Tall spike, many flowered
Flower	Pea flower shape, calyx tips entire, petals blue or violet
Fruit	Pod
Pollination	Entomophily
Seed dispersal	Autochory
Habitat	Grasslands, forests, gardens, orchards
Current status	Naturalised
Impacts	Decreases native plant diversity; allelopathic, potentially toxic to livestock, increases the nitrogen content of soil
Native range	Northern America
Global distribution	Asia-Tropical, Europe, Northern America

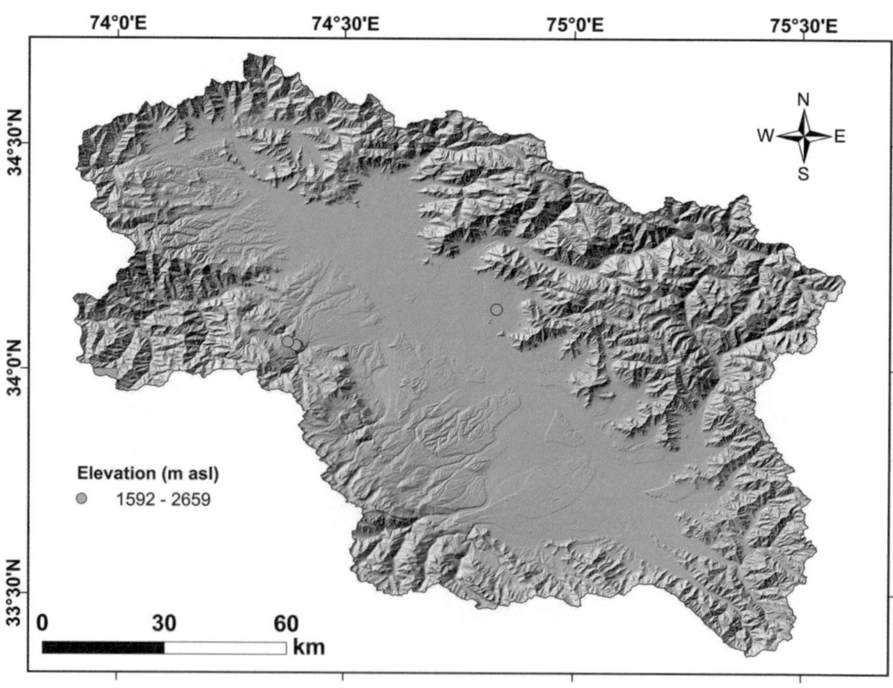

Lupinus polyphyllus: (**a**) Habit, (**b**) Leaf, (**c**) Inflorescence, (**d**) Flower, (**e**) Fruit

Malva parviflora L., Demonstr. Pl. 18 (1753).

Family	Malvaceae
English names	Least mallow, Cheese weed
Local name	'Panjeabi swatsal'
Habit	Annual herb, upto 90 cm tall
Stem	Erect or ascending, branched, sparsely hairy or glabrous
Leaves	Alternate, petiolate, leaf-blade subcordate to cordate, 5–7 lobed, base cordate, margin crenate, apex acute. Stipules lanceolate, persistent
Inflorescence	Axillary, compact
Flower	Pedicellate, epicalyx bracts linear to filiform, margin entire, sepals orbicular or deltoid green, petals white lilac or pinkish
Fruit	Schizocarp
Pollination	Entomophily
Seed dispersal	Autochory
Habitat	Agri-fields, roadsides, gardens, orchards, riparian areas
Current status	Invasive
Impacts	Decreases native plant diversity; reduces agricultural productivity, potentially toxic to livestock, reported to contain some toxic compound that causes skeletal muscle necrosis
Native range	Africa, Asia-Temperate, Europe
Global distribution	Africa, Asia-Temperate, Asia-Tropical, Europe, Northern America, Southern America, Australasia

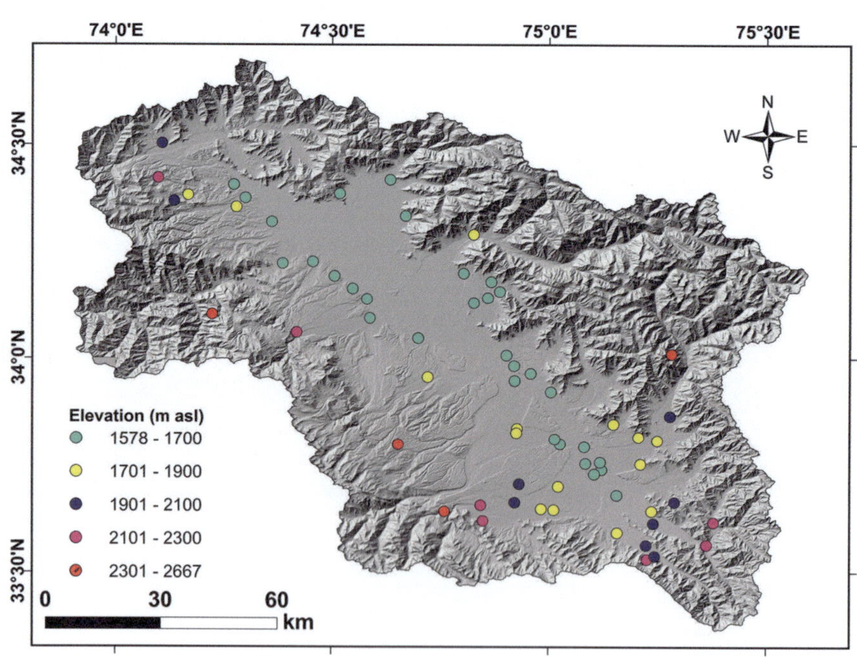

Malva parviflora: (**a**, **b**) Habit & Habitat, (**c**) Leaf, (**d**) Flower, (**e**) Inflorescence

Matricaria discoidea DC., Prodr. [A. P. de Candolle] 6: 50 (1838).

Family	Asteraceae
English names	Pineapple weed, Rayless chamomile
Local name	'Panjeab thool-i bober'
Habit	Annual herb, 5–40 cm tall
Stem	Erect or ascending, branched, glabrous or sparsely pubescent
Leaves	Alternate, sessile, pinnatisect, segments linear, both surfaces glabrous
Inflorescence	Capitula, single or corymbose, involucre cup-shaped, phyllaries in two rows
Flower	Ray florets absent, disc florets green, four-lobed
Fruit	Cypsela
Pollination	Entomophily
Seed dispersal	Anemochory, zoochory
Habitat	Grasslands, roadsides, gardens, orchards, agri-fields, forests, hillslopes
Current status	Invasive
Impacts	Reduces the forage value of pastures, decreases the native plant diversity and agricultural productivity
Native range	Asia-Temperate, Northern America
Global distribution	Asia-Tropical, Asia-Temperate, Europe, Northern America

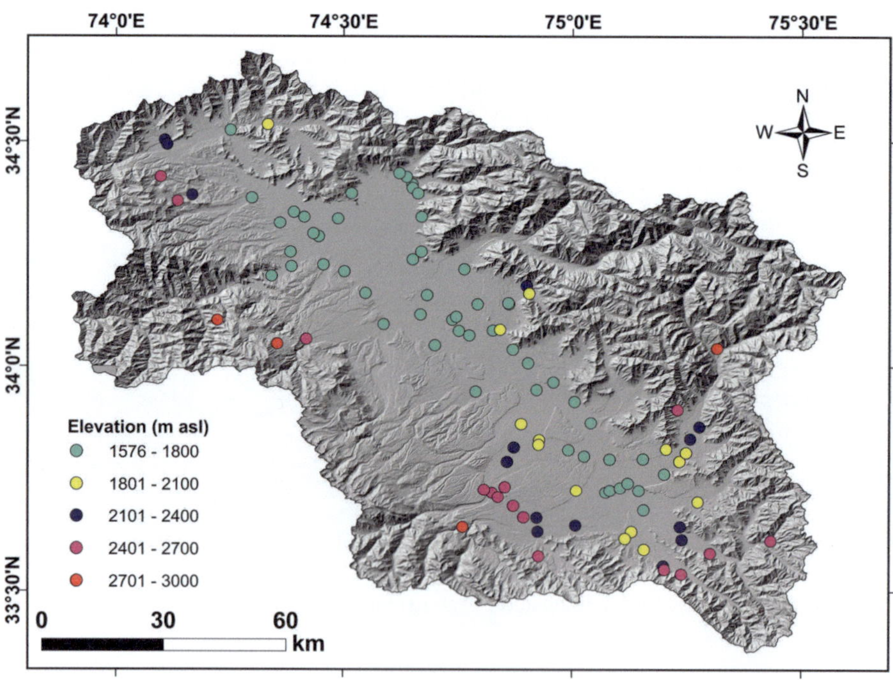

Matricaria discoidea: (**a**) Habit & Habitat, (**b**) Leaf, (**c**) Inflorescence

Medicago sativa L., Sp. Pl. 2: 778 (1753).

Family	Fabaceae
English name	Alfalfa
Local name	'Alfa gassi boeul'

Habit	Perennial herb, 30–60 cm tall
Stem	Erect or suberect, pubescent
Leaves	Trifoliate, petiolate, stipules lanceolate, leaflet obovate to ovate, hairy, margin dentate
Inflorescence	Peduncled raceme
Flower	Calyx tubular, teeth as long as tube, corolla bluish or violet
Fruit	Pod
Pollination	Entomophily
Seed dispersal	Anemochory, zoochory
Habitat	Grasslands, roadsides, gardens, orchards, agri-fields, forests, hillslopes

Current status	Invasive
Impacts	Decreases native plant diversity; reduces the quality and quantity of crop yield

Native range	Africa, Asia-Temperate, Asia-Tropical, Europe
Global distribution	Africa, Asia-Temperate, Asia-Tropical, Europe, Northern America, Southern America, Australasia

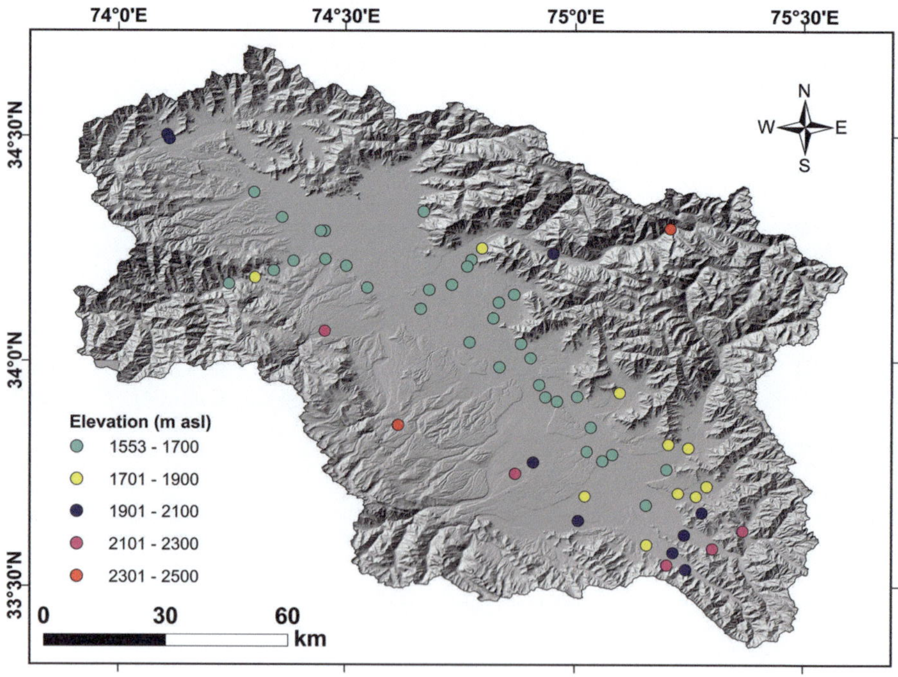

Medicago sativa: (**a**) Habit, (**b**) Leaf, (**c**) Inflorescence, (**d**) Flower

Mentha pulegium L., Sp. Pl. 2: 577 (1753).

Family	Lamiaceae
English name	European pennyroyal
Local name	'Yani'
Habit	Perennial herb, upto 90 cm tall
Stem	Prostrate to ascending, hirtellous, branched
Leaves	Lower stem leaves petiolate, upper ones subsessile, leaf-blade ovate to orbicular, margin entire to finely serrate, tip rounded
Inflorescence	Verticillaster
Flower	Calyx five-lobed, purple, dotted, hairy, corolla pale purple, two-lipped, upper lip notched, lower lip three-lobed
Fruit	Nutlet
Pollination	Entomophily
Seed dispersal	Autochory, hydrochory
Habitat	Riparian areas, roadsides, agri-fields, orchards
Current status	Invasive
Impacts	Decreases native plant diversity; reduces the quality and quantity of crop yield, declines the forage value of pastures, reported to be harmful to milch cows
Native range	Africa, Asia-Temperate, Europe
Global distribution	Africa, Asia-Temperate, Asia-Tropical, Europe, Australasia, Northern America, Pacific, Southern America

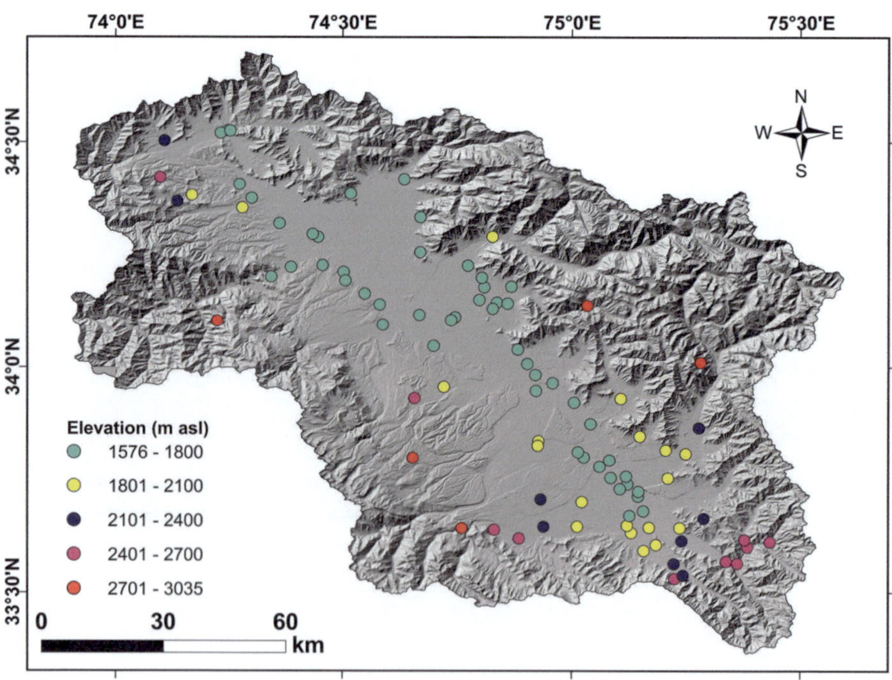

Mentha pulegium: (**a**) Habitat, (**b**) Habit, (**c**) Inflorescence

Muscari neglectum Ten., Ann. Civili Regno Due Sicilie 26: 49 (1841).

Family	Asparagaceae
English name	Grape hyacinth
Local name	'Dachh-i-hyacinth'
Habit	Perennial geophyte
Stem	Scapose, 20–70 cm tall
Leaves	Rosette, long, linear, arise from the ovoid bulb
Inflorescence	Raceme, 20–40 flowered
Flower	Pedicellate, urn-shaped, dark blue, fragrant
Fruit	Loculicidal capsule
Pollination	Entomophily
Seed dispersal	Autochory
Habitat	Gardens, orchards, grasslands, forests
Current status	Casual
Impacts	Decreases the native plant diversity, bulbs are reported to be poisonous
Native range	Africa, Asia-Temperate, Asia-Tropical, Europe
Global distribution	Africa, Asia-Temperate, Asia-Tropical, Europe, Northern America, Southern America

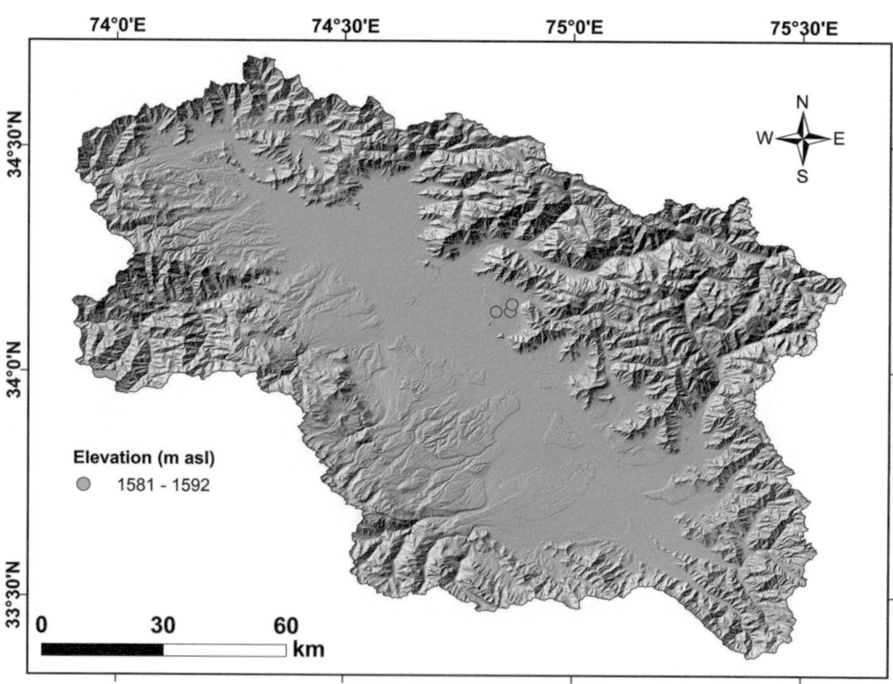

Muscari neglectum: (**a**) Habitat, (**b**) Habit (Inset: Bulb), (**c**) Inflorescence

Myriophyllum aquaticum (Vell.) Verdc., Kew Bull. 28 (1): 36 (1973).

Family	Halogragaceae
English name	Parrot's feather
Local name	'Toti teer'

Habit	Perennial aquatic herb, 100–200 cm long
Stem	Simple or sparsely branched at base
Leaves	Alternate, pinnatisect, papillose, segments 8–16 on each side, linear in shape, base dilated, apex mucronulate
Inflorescence	Solitary axillary
Flower	Unisexual, female flowers in lower leaf axils, male flowers in upper leaf axils
Fruit	Schizocarp
Pollination	No sexual reproduction as male plants absent
Habitat	Lakes, ponds, streams, rivers, canals, riparian areas

Current status	Invasive
Impacts	Reduces the quality of water, causes disruption in flow of irrigation water, interferes in water navigation, causes loss to fishery and recreation; reported to cause blockage of waterways and lakes

Native range	Southern America
Global distribution	Africa, Asia-Temperate, Asia-Tropical, Northern America, Pacific, Southern America, Australasia

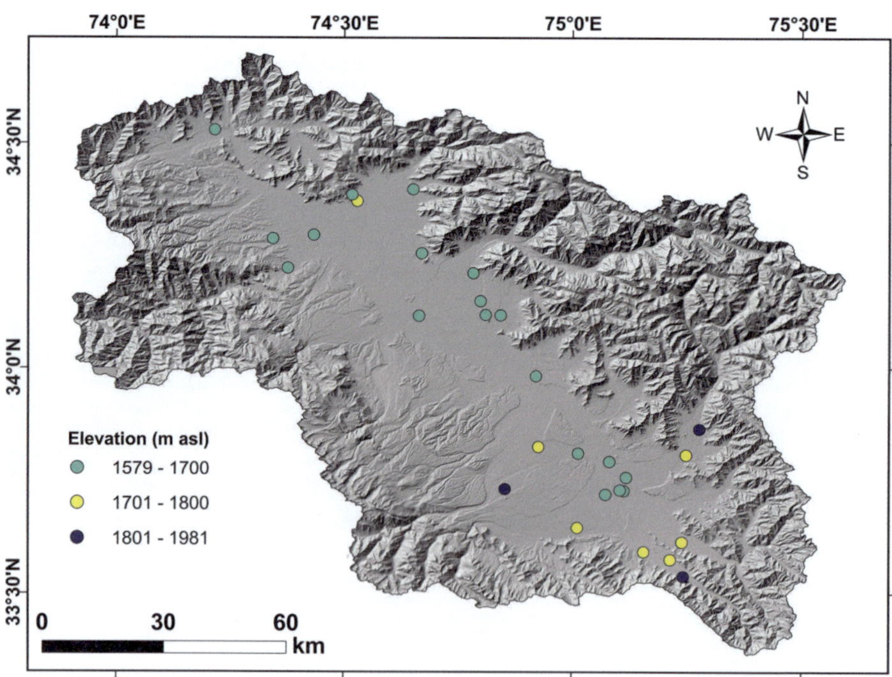

Myriophyllum aquaticum: (**a**) Habitat, (**b**) Leaf, (**c**) Inflorescence

Narcissus poeticus L., Sp. Pl. 1: 289 (1753).

Family	Amaryllidaceae
English names	Poet's daffodil, Pheasant's eye
Local name	'Masval'
Habit	Perennial herb, bulbs ovoid, tunicated, brown
Stem	Scapose, upto 80 cm tall
Leaves	Arise from bulb, leaf-blade flat, green to glaucous
Inflorescence	One flowered, spathe pale brown
Flower	Fragrant, tepals overlapping, white, ovate to round, corona yellow, cup-shaped, with red crenulate margin
Flower	Capsule
Pollination	Entomophily
Seed dispersal	Anemochory
Habitat	Gardens, roadsides, forests, orchards
Current status	Naturalised
Native range	Europe
Global distribution	Asia-Temperate, Europe, Northern America, Southern America, Australasia

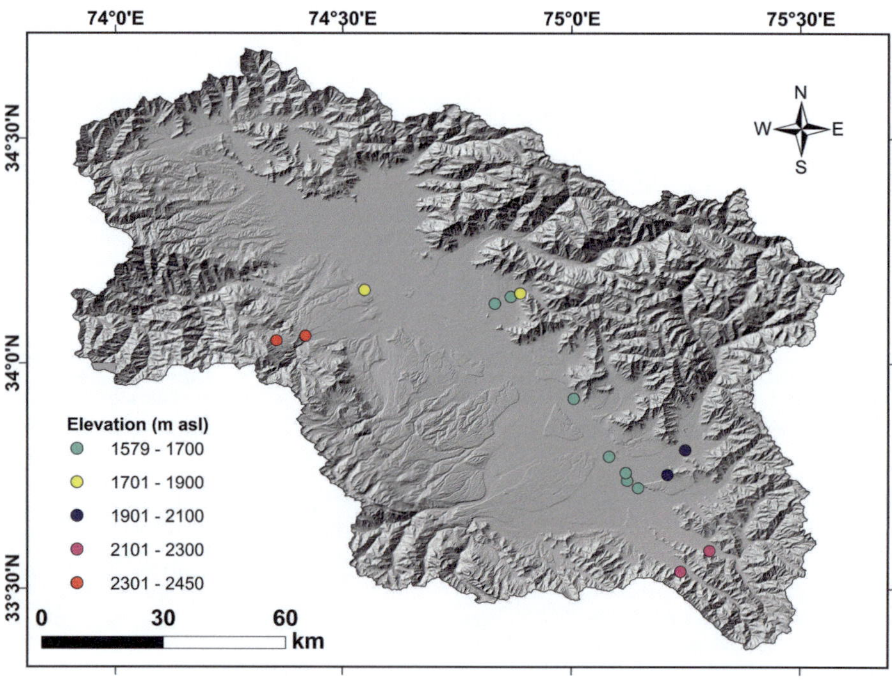

***Narcissus poeticus*:** (**a, b**) Habit & Habitat, (**c**) Flower, (**d**) Fruit

Narcissus pseudonarcissus L., Sp. Pl. 1: 289 (1753).

Family	Amaryllidaceae
English name	Wild daffodil
Local name	'Speaker yimbirzal'
Habit	Perennial herb, bulbs ovoid, tunicated, pale brown
Stem	Scapose, 25–30 cm tall
Leaves	Arise from bulb, strap like, grey green
Inflorescence	Single flower is produced at the top of the flattened flower stalk
Flower	Fragrant, light yellow perianth, dark yellow long trumpet (corona), tip of corona wavy
Fruit	Capsule
Pollination	Entomophily
Seed dispersal	Anemochory
Habitat	Gardens, roadsides, forests, orchards
Current status	Naturalised
Native range	Africa, Asia-Temperate, Europe
Global distribution	Africa, Asia-Temperate, Europe

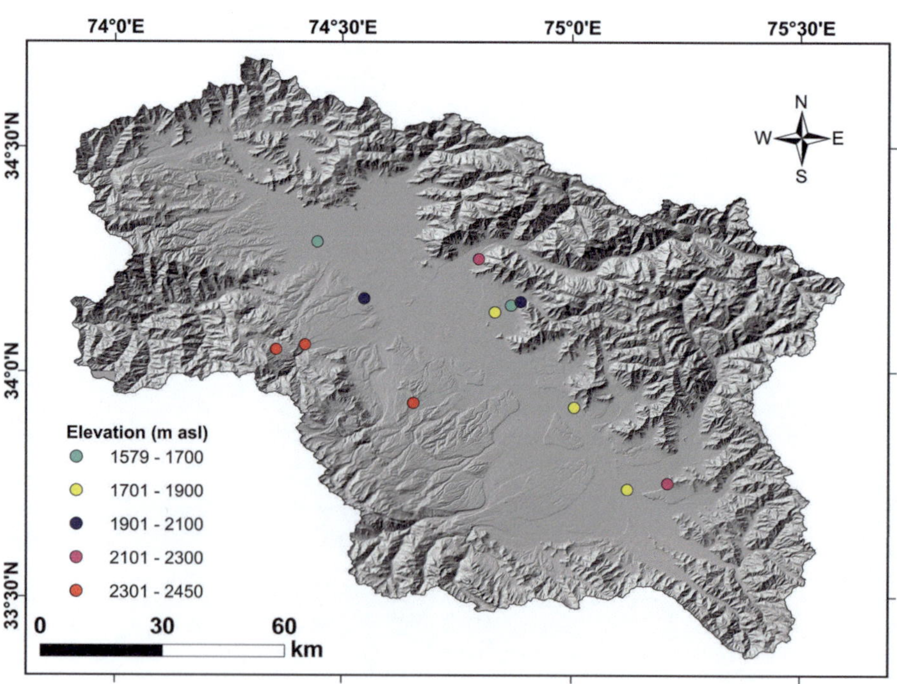

***Narcissus pseudonarcissus*:** (**a**, **b**) Habit & Habitat, (**c**, **d**) Flower

Narcissus tazetta L., Sp. Pl. 1: 290 (1753).

Family	Amaryllidaceae
English name	Daffodil
Local name	'Yimbirzal'
Habit	Perennial herb, bulb subterranean
Stem	Scapose, 24–30 cm tall
Leaves	Arise from bulb, narrow strap like, 15–70 cm long, grow upward and then droop down
Inflorescence	One to many flowers per stalk
Flower	Fragrant, tepals white or cream, corona cup-shaped, yellow
Fruit	Capsule
Pollination	Entomophily
Seed dispersal	Anemochory
Habitat	Gardens, roadsides, forests, orchards
Current status	Naturalised
Native range	Africa, Asia-Temperate, Europe
Global distribution	Africa, Asia-Temperate, Europe

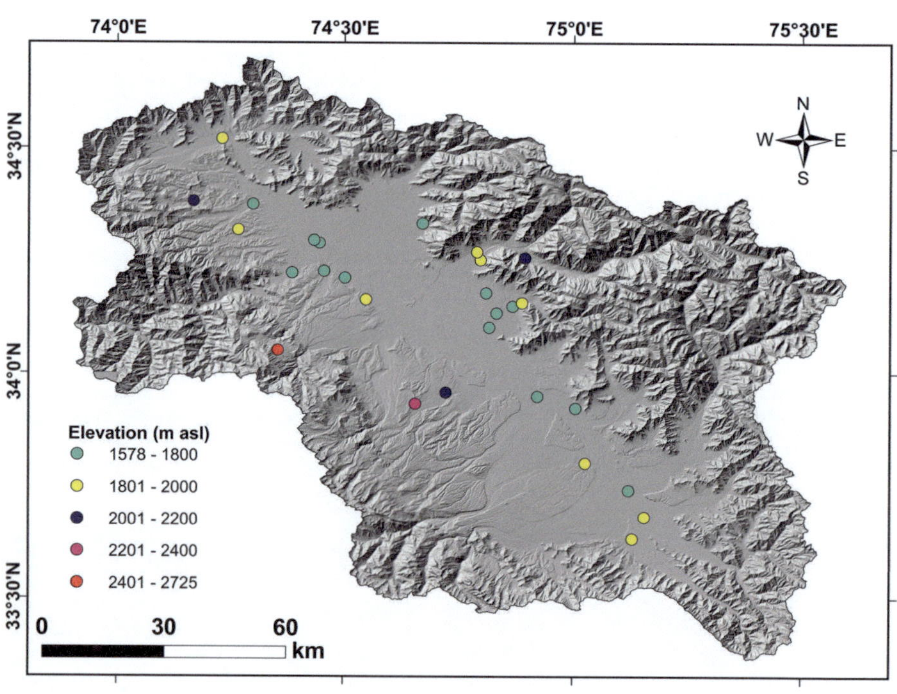

Narcissus tazetta: (**a**) Habitat, (**b**) Habit, (**c**) Flower, (**d**) Fruit

Nymphaea mexicana Zucc., Abh. Math.-Phys. Cl. Königl. Bayer. Akad. Wiss. 1: 365. (1832).

Family	Nymphaeaceae
English name	Mexican water-lily
Local name	'Lyodur bum-posh'
Habit	Perennial rhizomatous herb
Stem	Rhizomes unbranched, erect, cylindric, stolons spongy
Leaves	Petiolate, leaf-blade elliptic to nearly round, base cordate, apex obtuse, margin entire or sinuate
Inflorescence	Solitary
Flower	Floating, attractive, sepals yellowish green, petals yellow
Fruit	Berry-like
Pollination	Entomophily
Seed dispersal	Hydrochory, ornithochory
Habitat	Lakes, ponds, streams, rivers, canals
Current status	Invasive
Impacts	Restricts water movement, decreases recreational value of water bodies, declines light availability to aquatic vegetation and reduces quality of water; decreases the native plant diversity
Native range	Northern America
Global distribution	Africa, Asia-Tropical, Europe, Northern America

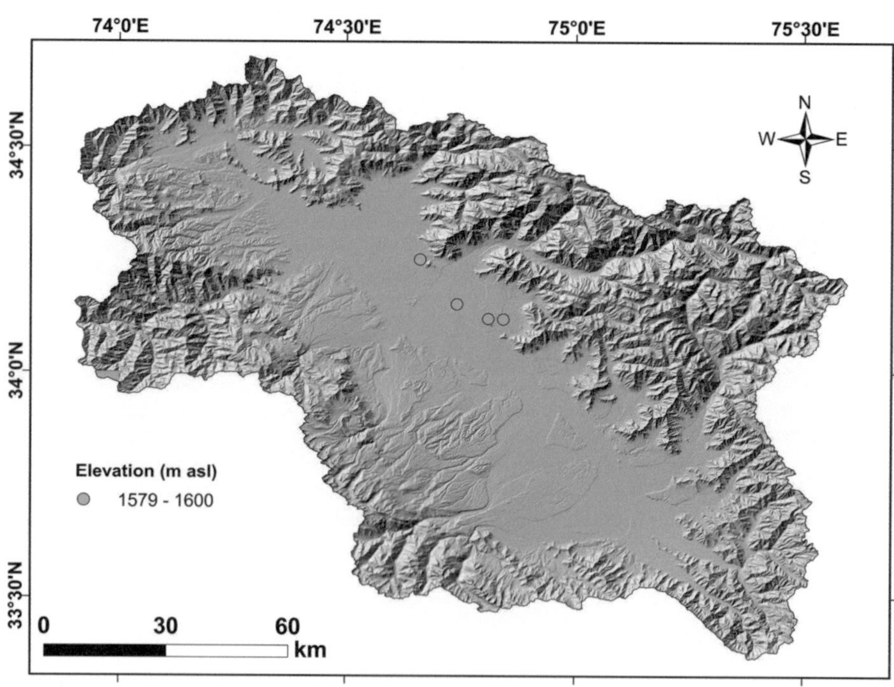

***Nymphaea mexicana*:** (**a**) Habitat, (**b**) Habit, (**c**) Leaf, (**d**, **e**) Flower

Oenothera glazioviana Micheli., Fl. Bras. (Martius) 13(2): 178 (1875).

Family	Onagraceae
English name	Large-flower evening primrose
Local name	'Sham-i sundar'
Habit	Biennial herb, 200–250 cm tall
Stem	Erect, simple or branched, hairy, hairs with purplish base
Leaves	Rosette leaves lanceolate to oblanceolate, cauline leaves alternate, leaf-blade elliptic to lanceolate, narrowed to the petiole, uppermost leaves sessile, margin wavy, toothed to serrulate
Inflorescence	Spike
Flower	Flower tube long, sepals red striped along the midrib, petals yellow, broadly obcordate
Fruit	Capsule
Pollination	Entomophily, autogamy
Seed dispersal	Ornithochory, autochory
Habitat	Agri-fields, grasslands, roadsides, gardens
Current status	Invasive
Impacts	Decreases native plant diversity; reduces the quality and quantity of crop yield
Native range	Southern America
Global distribution	Africa, Asia-Temperate, Asia-Tropical, Europe, Northern America, Southern America

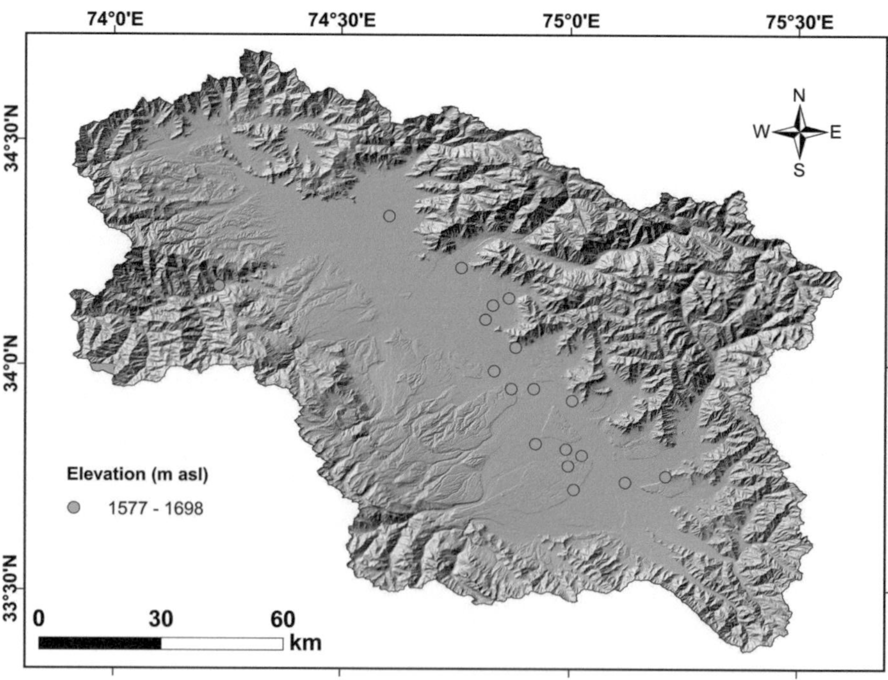

***Oenothera glazioviana*:** (**a**) Habit, (**b**) Leaf, (**c**) Inflorescence, (**d**) Flower, (**e**) Fruit

Onopordum acanthium L., Sp. Pl. 2: 827 (1753).

Family	Asteraceae
English name	Cotton thistle
Local name	'Dod-i kond'

Habit	Biennial or perennial herb, 200–250 cm tall
Stem	Erect, branched above, glabrous, spiny
Leaves	Alternate, deeply lobed with long stiff spines along the margins, basal leaves elliptic to broadly ovate, upper and middle cauline leaves sessile, narrowly elliptic to oblanceolate
Inflorescence	Capitula solitary, involucre globose to ovoid
Flower	Corolla purplish red to pink
Fruit	Achene
Pollination	Anemophily
Seed dispersal	Anemochory
Habitat	Grasslands, roadsides, hill slopes, gardens, agri-fields

Current status	Invasive
Impacts	Declines the forage value of pastures, reduces the quality and quantity of crop yield; sharp prickles cause injury to humans and livestock, decreases the native plant diversity

Native range	Africa, Asia-Temperate, Asia-Tropical, Europe
Global distribution	Africa, Asia-Temperate, Asia-Tropical, Europe, Northern America, Southern America, Australasia

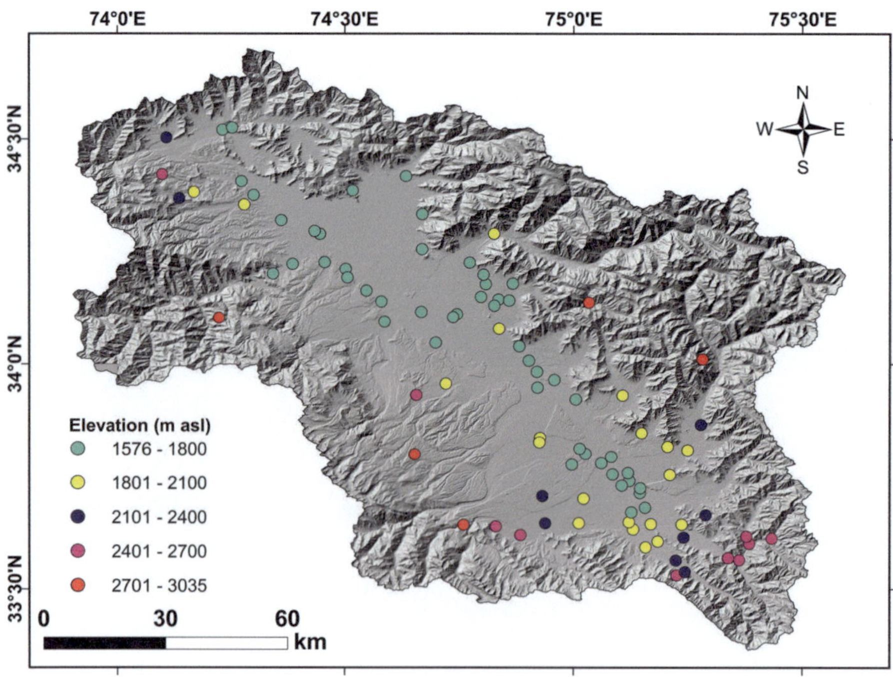

***Onopordum acanthium*: (a)** Habitat, **(b)** Habit, **(c)** Inflorescence

Ornithogalum umbellatum L., Sp. Pl. 1: 307 (1753).

Family	Asparagaceae
English names	Garden star-of-Bethlehem, Grass lily
Local name	'Gassi lily'
Habit	Perennial bulbous herb
Stem	Flowering stem 10–30 cm tall, glabrous
Leaves	Arise in tufts from base, leaf-blade strap like, long, smooth, white to light green mid-rib on upper surface
Inflorescence	Corymb
Flower	Bisexual, tepals 6, white, green stripes on outer surface, stamens in 2 rows
Fruit	Capsule
Pollination	Entomophily
Seed dispersal	Autochory
Habitat	Gardens, orchards, grasslands, forests
Current status	Naturalised
Impacts	Decreases native plant diversity; reported to be toxic to humans and animals, poisonous symptoms include diarrhoea, vomiting, nausea, shortness of breath, as well as pain, burning, and swelling of lips, tongue, and throat; sometimes protracted contact may cause skin irritation
Native range	Africa, Asia-Temperate, Europe
Global distribution	Africa, Asia-Temperate, Asia-Tropical, Europe, Northern America, Australasia

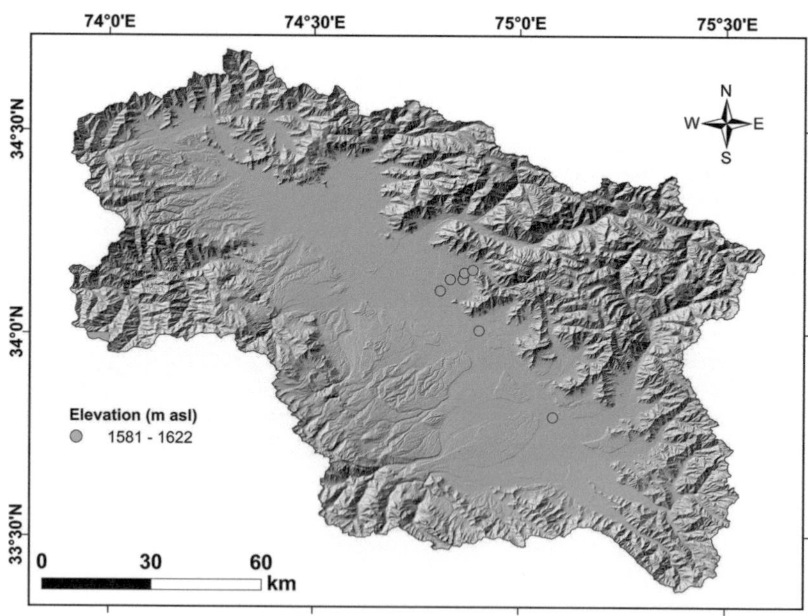

Ornithogalum umbellatum: (**a**) Habit, (**b**) Inflorescence, (**c**) Flower, (**d**) Perianth, (**e**) Bulb

Oxalis corniculata L., Sp. Pl. 1: 435 (1753).

Family	Oxalidaceae
English name	Creeping wood sorrel
Local name	'Lyodur tchok-i tchin'
Habit	Annual or perennial herb, 10–25 cm tall
Stem	Creeping or ascending, branched, purplish, pubescent
Leaves	Alternate, petiolate, trifoliate, leaflet-blade obcordate, pubescent on both surfaces, margin entire. Stipules small, glabrous
Inflorescence	Umbelliform cymes, 2–7 flowered
Flower	Sepals ovate to lanceolate, hairy, petals yellow, slightly emarginate
Fruit	Capsule
Pollination	Entomophily
Seed dispersal	Anemochory
Habitat	Grasslands, roadsides, gardens, orchards, agri-fields, forests, hillslopes
Current status	Invasive
Impacts	Decreases native plant diversity; reduces crop yield, reported to be poisonous to livestock, muscle shaking, stumbling, heavy breathing, losing control of the hindquarters and slumping to the ground, and a slimy discharge from the nostrils are all signs of poisoning
Native range	Asia-Temperate, Northern America
Global distribution	Africa, Asia-Temperate, Asia-Tropical, Europe, Northern America, Pacific, Southern America, Australasia

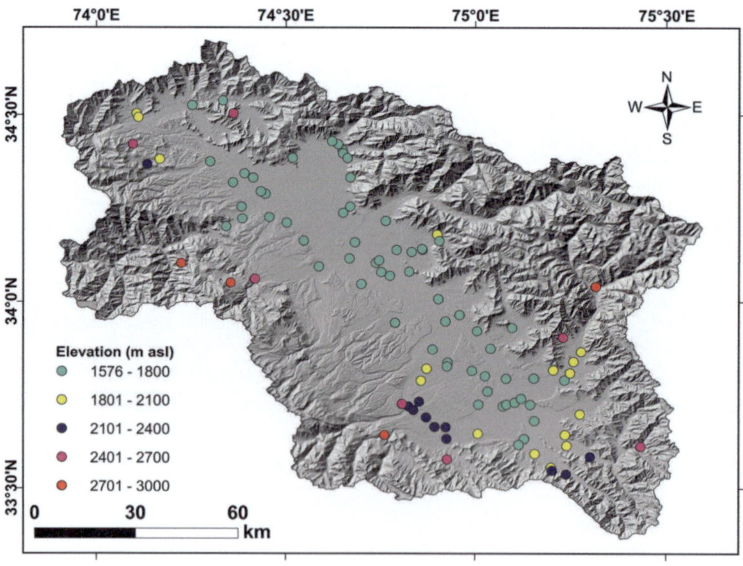

Oxalis corniculata: (**a**) Habit, (**b**) Flower, (**c**) Calyx

Oxalis debilis Kunth, Nov. Gen. Sp. [H.B.K.] v. 236 (1822).

Family	Oxalidaceae
English name	Large-flowered pink sorrel
Local name	'Goleab tchok-i tchin'
Habit	Perennial herb, bulbous
Stem	Flowering stem upto 30 cm tall
Leaves	Petiolate, trifoliate, leaf-blade obcordate, sparsely hairy on both surfaces, apex notched, stipules connate
Inflorescence	Cyme
Flower	Pedicellate, calyx greenish, margin hyaline, corolla pinkish or rose purple, clawed
Fruit	Capsule
Pollination	Entomophily
Seed dispersal	Anemochory
Habitat	Agri-fields, roadsides, gardens, orchards, grasslands, hillslopes
Current status	Naturalised
Impacts	Reduces crop yield, potentially allelopathic, decreases native plant diversity; reported to be poisonous to livestock if consumed in large quantity
Native range	Southern America
Global distribution	Africa, Asia-Temperate, Asia-Tropical, Europe, Northern America, Southern America, Australasia

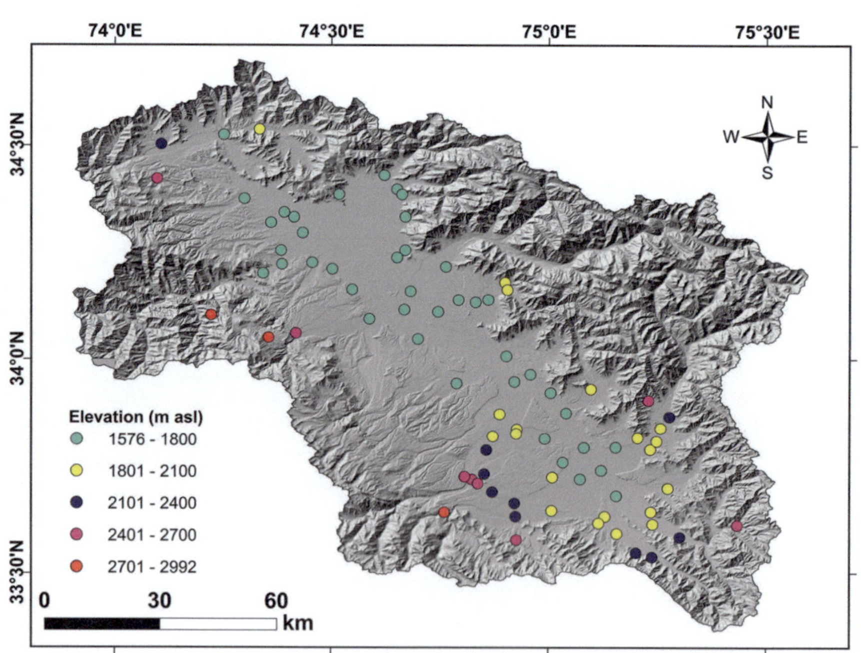

***Oxalis debilis*:** (**a**, **b**) Habit & Habitat, (**c**) Flower (Inflorescence)

Papaver dubium L., Sp. Pl. 2: 1196 (1753).

Family	Papaveraceae
English name	Long-head poppy
Local name	'Thod Gulaali'

Habit	Annual herb, 30–70 cm tall
Stem	Erect, branched, glabrous to sparsely hispid
Leaves	Alternate, pinnatifid, segments lobed, apex acute
Inflorescence	Solitary terminal, peduncle hairy
Flower	Bisexual, sepals 2, glabrous or bristly, petals 4, suborbicular to obovate, reddish or pale scarlet with basal dark spots, stamens as long as ovary
Fruit	Capsule, oblong in shape
Pollination	Entomophily
Seed dispersal	Anemochory
Habitat	Grasslands, agri-fields, roadsides, gardens, orchards, forests

Current status	Naturalised
Impacts	Decreases native plant diversity; reduces crop yield

Native range	Africa, Asia-Temperate, Asia-Tropical, Europe
Global distribution	Africa, Asia-Temperate, Asia-Tropical, Europe, Northern America

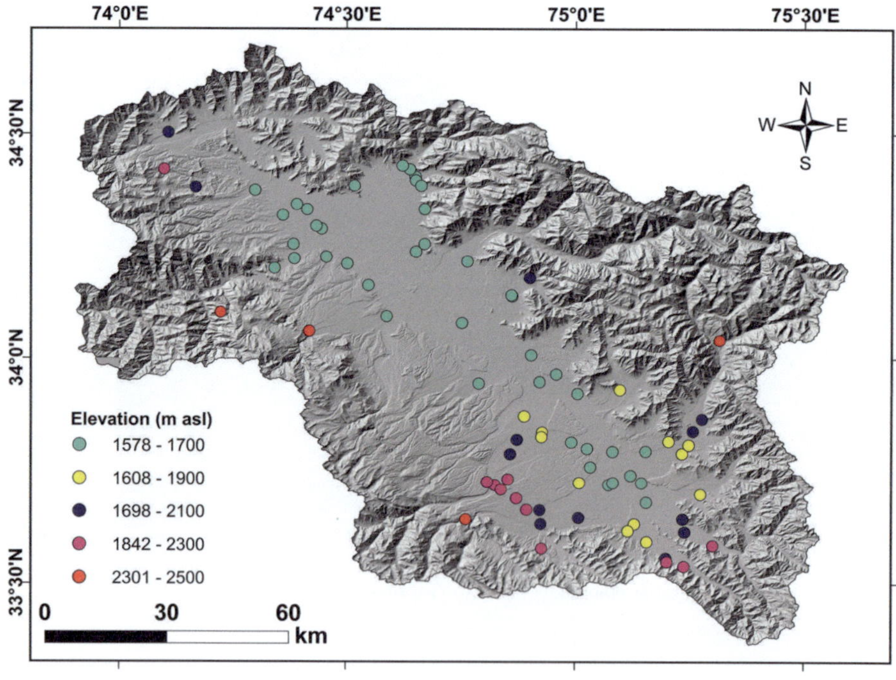

Papaver dubium: (**a**) Habitat, (**b**) Leaf, (**c**) Floral bud, (**d**, **e**) Flower, (**f**) Fruit

Papaver rhoeas L., Sp. Pl. 1: 507 (1753).

Family	Papaveraceae
English name	Common poppy
Local name	'Aam Gulaali'
Habit	Annual herb, 30–100 cm tall
Stem	Erect, branched, hairy
Leaves	Alternate, pinnatisect, lower leaves petiolate, upper sessile, lobes linear to lanceolate, apex acute, margin entire or toothed
Inflorescence	Solitary, peduncle 10–20 cm long
Flower	Bisexual, sepals 2, petals 4, obovate to orbicular, brick red or scarlet red, stamens longer than ovary
Fruit	Capsule, oblong to ellipsoid in shape
Pollination	Entomophily
Seed dispersal	Anemochory
Habitat	Grasslands, agri-fields, roadsides, gardens, orchards, forests
Current status	Naturalised
Impacts	Decreases native plant diversity; reduces crop yield
Native range	Africa, Asia-Tropical, Europe
Global distribution	Africa, Asia-Tropical, Europe, Australasia

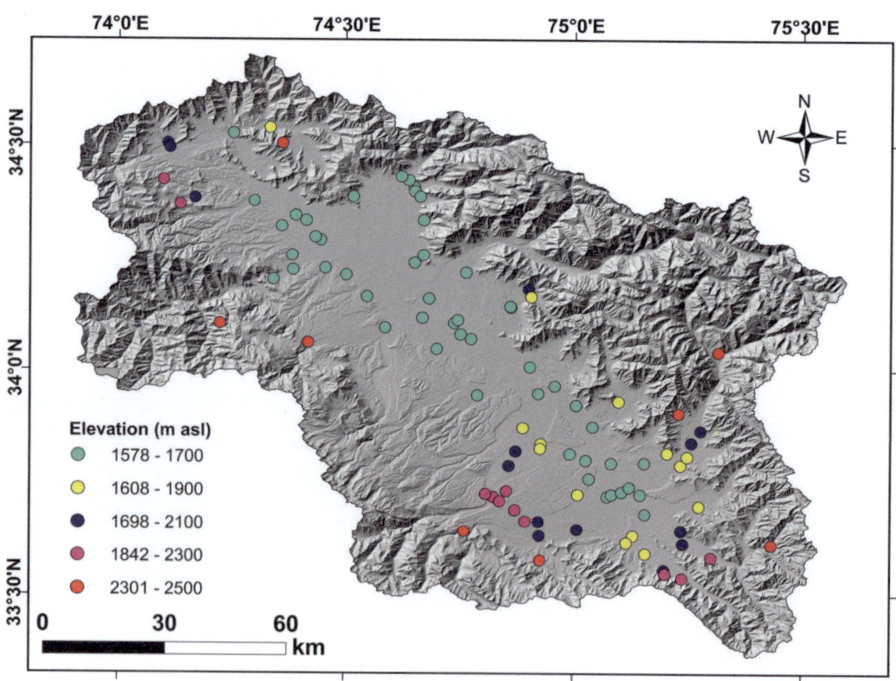

Papaver rhoeas: (**a**) Habitat, (**b**) Leaf, (**c**) Floral bud, (**d**) Flower, (**e**) Fruit

Parthenium hysterophorus L., Sp. Pl. 2: 988 (1753).

Family	Asteraceae
English name	Carrot grass
Local name	'Parim gazrii-gassi'
Habit	Annual herb, upto 200 cm tall
Stem	Erect, branched, sparsely hairy
Leaves	Alternate, petiolate, leaf-blade ovate to elliptic, pinnately lobed, pale green, pubescent
Inflorescence	Capitula with five distinct corners, involucre cup-shaped
Flower	Ray florets white, distally notched, disc florets with lobed corolla
Fruit	Achene
Pollination	Entomophily
Seed dispersal	Anemochory, zoochory
Habitat	Agri-fields, roadsides, gardens, orchards
Current status	Invasive
Impacts	Decreases native plant diversity; reduces the quantity and quality of crop yield; reported to be highly allergic to humans and animals and the symptoms include hay fever, body pain dermatitis; also acts as a secondary host to many crop plant diseases
Native range	Northern America, Southern America
Global distribution	Africa, Asia-Temperate, Asia-Tropical, Northern America, Pacific, Southern America, Australasia

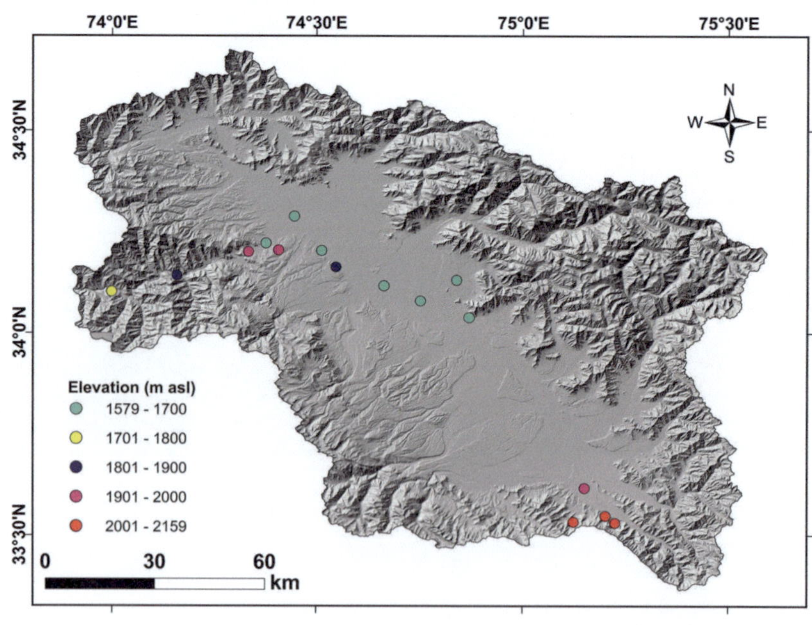

Parthenium hysterophorus: (**a**) Habitat, (**b**) Habit, (**c**) Leaf, (**d**) Inflorescence

Populus deltoides W. Bartram ex Marshall. Arbust. Amer. 106 (1785).

Family	Salicaceae
English name	Eastern cottonwood
Local name	'Roosi phrast'
Habit	Large tree, 20–60 m tall
Stem	Erect, branched, bark slivery white when young, dark grey when old, twigs greyish yellow
Leaves	Petiolate, leaf-blade deltoid, surface glabrous, base truncate, apex acuminate, margin serrate
Inflorescence	Catkin, male catkins reddish purple, female catkins green
Flower	Saucer-shaped
Fruit	Capsule
Pollination	Entomophily
Seed dispersal	Anemochory, hydrochory
Habitat	Riparian areas, roadsides, orchards
Current status	Naturalised
Impacts	Decreases native plant diversity; fluffy cotton-covered abortive seeds cause many respiratory diseases like bad cold, cough, fever, and sore throat
Native range	Northern America
Global distribution	Asia-Temperate, Asia-Tropical, Europe, Northern America

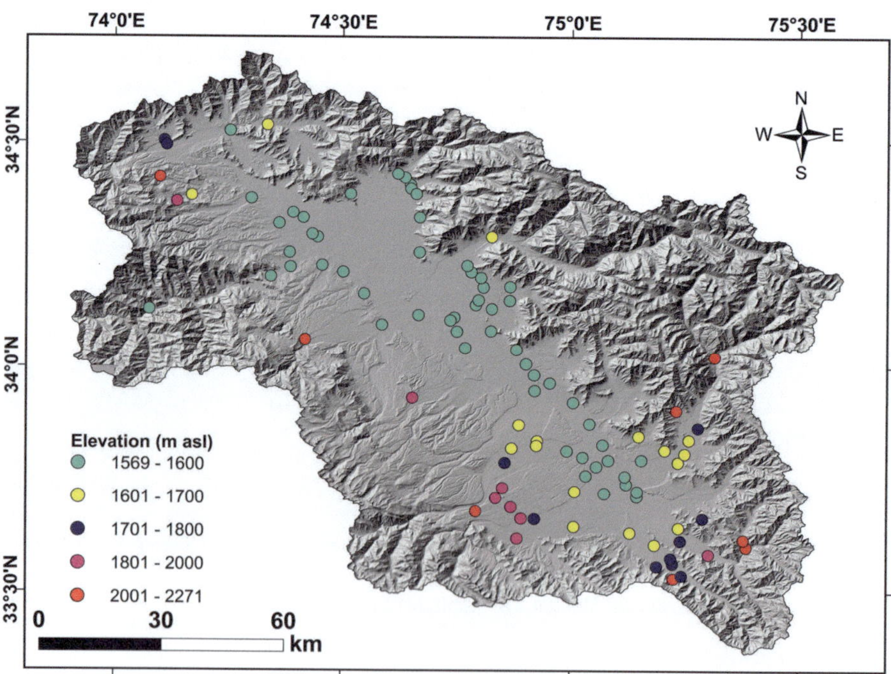

Populus deltoides: (**a**) Habit & Habitat, (**b**) Leaf, (**c**, **d**) Fruit

Quercus robur L., Sp. Pl. 2: 996 (1753).

Family	Fagaceae
English name	English oak
Local name	'Jamal goti'

Habit	Deciduous tree, upto 50 m tall
Stem	Trunk, solid, strong, lenticellate
Leaves	Alternate, pinnatifid, leaf-blade ovate, upper surface dark green, lower surface pale green, pubescent at veins, margin sinuate
Inflorescence	Catkin
Flower	Male flowers in lax (loose) catkins, female flowers in stout catkins, perianth segments lanceolate, tomentose
Fruit	Nut (Acorn)
Pollination	Anemophily
Seed dispersal	Ornithochory, zoochory
Habitat	Forests, hillslopes

Current status	Naturalised
Impacts	Decreases the light availability in forest ecosystems, produces large amount of difficult to decompose litter; reported to increase the tannin content in soil and disturbs the nutrient cycling; decreases the native plant diversity

Native range	Europe
Global distribution	Asia-Temperate, Europe

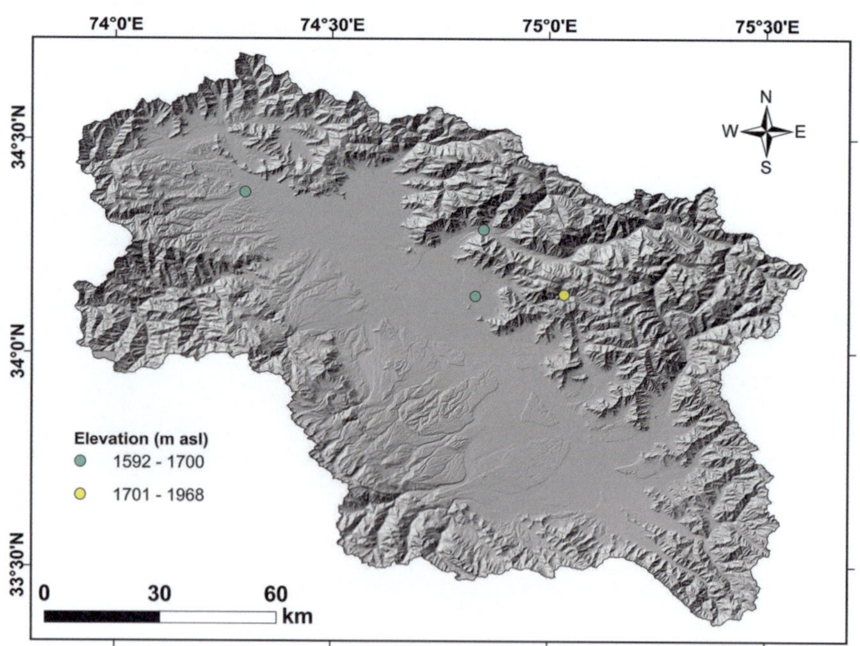

Quercus robur: (**a**) Habit, (**b**) Sapling, (**c, d**) Fruit

Ranunculus bulbosus L., Sp. Pl. 1: 554 (1753).

Synonym	*Ranunculus laetus* Salisb
Family	Ranunculaceae
English names	Bulbous buttercup, Cheerful buttercup
Local name	'Mond-dar buttercup'
Habit	Perennial herb, upto 90 cm tall
Stem	Erect, branched, sparsely hairy or glabrous
Leaves	Lower stem leaves petiolate, 3–5 lobed, each lobe obovate in outline, margin serrate, surface pubescent, cauline leaves smaller and less divided
Inflorescence	Solitary terminal and axillary
Flower	Pedicellate, sepals yellowish, adaxial surface hairy, petals 5, obovate, bright yellow
Fruit	Achene
Pollination	Entomophily
Seed dispersal	Hydrochory
Habitat	Grasslands, roadsides, gardens, orchards, agri-fields, forests, hillslopes
Current status	Invasive
Impacts	Decreases native plant diversity, reduces quantity and quality of crop yield; reported to cause skin blisters in humans and poisonous to livestock
Native range	Africa, Asia-Temperate, Europe
Global distribution	Africa, Asia-Temperate, Europe, Northern America, Southern America, Australasia

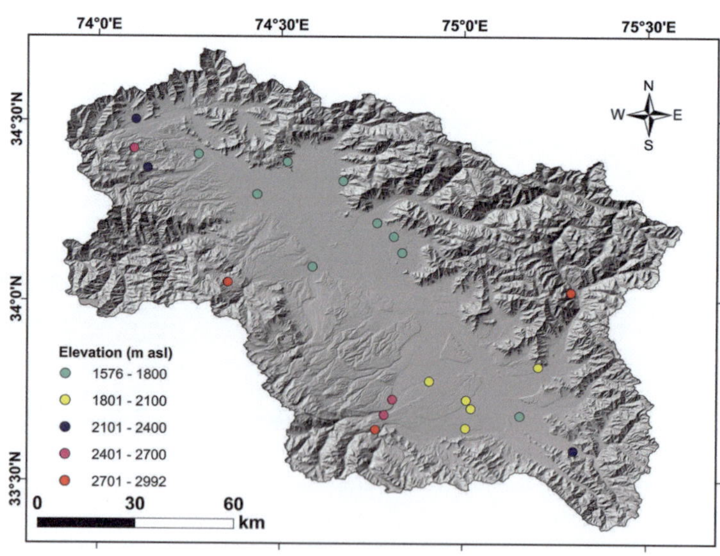

Ranunculus bulbosus: (**a**) Habit, (**b**) Leaf, (**c**) Flower, (**d**) Fruit

Ranunculus repens L., Sp. Pl. 1: 554 (1753).

Family	Ranunculaceae
English name	Creeping buttercup
Local name	'Pather buttercup'
Habit	Perennial herb, 10–50 cm tall
Stem	Creeping or ascending, branched, pubescent
Leaves	Basal leaves petiolate, lamina ternate, central lobe stalked, rhombic, glabrous or sparsely pubescent on adaxial surface and prominently pubescent on abaxial surface, base cuneate, margin dentate, cauline leaves smaller, subsessile to sessile upwards
Inflorescence	Inflorescence monochasium, 2 to several flowered, 2–3 cm in diameter, pedicel 2–10 cm long, pubescent
Flower	Sepals 5, elliptic-ovate, sparsely pubescent abaxially, petals golden yellow, 2–3 seriate, obovate, finely veined, nectary at the base, apex rounded
Fruit	Achene
pollination	Entomophily, autogamy
Seed dispersal	Ornithochory, zoochory, hydrochory
Habitat	Riparian areas, grasslands
Current status	Naturalised
Impacts	Decreases native plant diversity; reduces the forage value of pastures, reported to deplete potassium in soil, accumulates sediments in rhizomes and toxic to livestock
Native range	Africa, Asia-Temperate, Europe
Global distribution	Africa, Asia-Temperate, Asia-Tropical, Europe

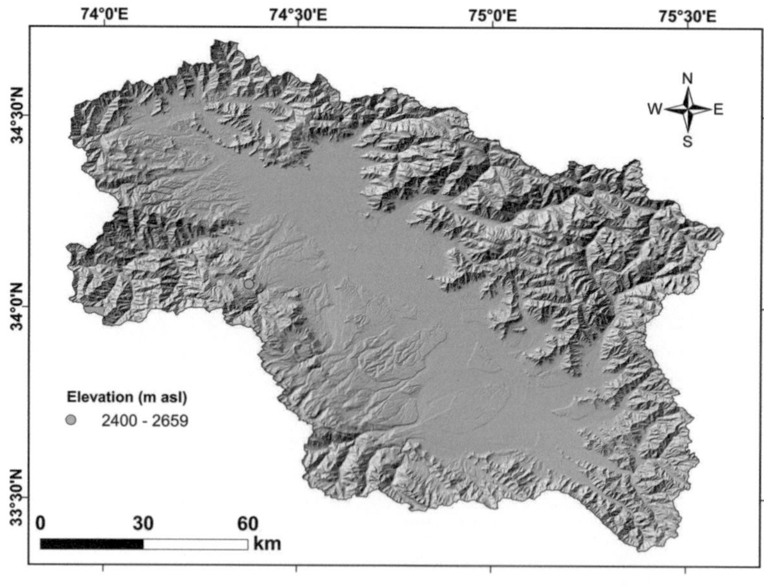

Ranunculus repens: **(a)** Habit, **(b)** Leaf, **(c)** Flower, **(d)** Fruit

Ricinus communis L., Sp. Pl. 2: 1007 (1753).

Family	Euphorbiaceae
English name	Castor bean
Local name	'Castor tili kul'
Habit	Perennial robust herb, 200–500 cm tall
Stem	Erect, branched, purplish, glabrous
Leaves	Alternate, petiolate, leaf-blade palmately compound, 7–11 lobed, margin serrate, stipules connate
Inflorescence	Panicle
Flower	Lower stem nodes possess clusters of staminate flowers, upper stem nodes possess pistillate flowers
Fruit	Spiny capsule
Pollination	Anemophily
Seed dispersal	Ornithochory, zoochory, hydrochory
Habitat	Agri-fields, roadsides, grasslands
Current status	Casual
Impacts	Reduces the forage value of pastures, reported to be extremely poisonous to humans and animals; decreases the native plant diversity, causes respiratory allergic problems in humans
Native range	North East Africa
Global distribution	Africa, Asia-Temperate, Asia-Tropical, Europe, Northern America, Southern America, Australasia

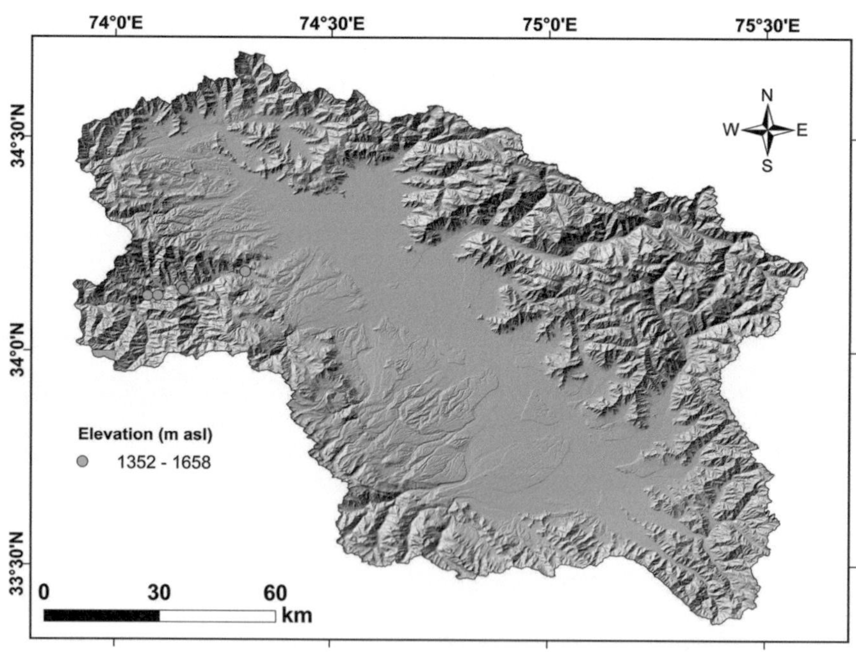

***Ricinus communis*: (a, b)** Habit & Habitat, **(c, e)** Flower, **(d, f)** Fruit

Robinia pseudoacacia L., Sp. Pl. 2: 722 (1753).

Family	Fabaceae
English name	Black locust tree
Local name	'Kyikar kul'
Habit	Deciduous tree, 15–25 m tall
Stem	Erect, branched, bark dark brown, fissured
Leaves	Alternate, pinnately compound, leaflets 2–12 pairs, leaflet ovate, tip notched, margin entire, pair of short thorns at the base of each leaf
Inflorescence	Raceme, pendulous, many flowered, fragrant
Flower	Calyx campanulate, pubescent, reddish purple, corolla white, inside with yellow spots
Fruit	Pod with reddish brown stripes
Pollination	Entomophily
Seed dispersal	Anemochory, zoochory
Habitat	Agri-fields, roadsides, gardens, grasslands, orchards, forests, hillslopes
Current status	Invasive
Impacts	Decreases native plant diversity; causes pollen allergy in humans, reported to alter the nitrogen content of soil
Native range	Northern America
Global distribution	Africa, Asia-Temperate, Asia-Tropical, Europe, Northern America, Southern America, Australasia

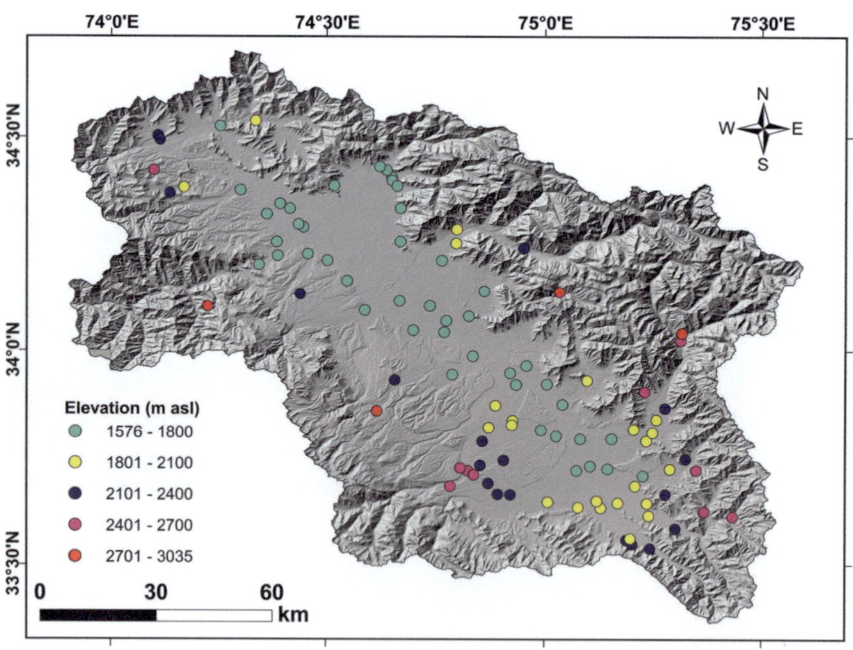

Robinia pseudoacacia: (**a**) Habit, (**b**) Spiny Stem, (**c**) Floral bud, (**d**) Inflorescence, (**e**) Fruit

Rubus ulmifolius Schott, Isis. (Oken) 1818, (Heft 5): 821 (1818).

Family   Rosaceae
English name   Elm-leaf blackberry
Local name   'Chhaanchh'

Habit   Scrambling shrub, upto 300 cm tall
Stem   Erect, angled, hairy, prickles present, slightly reddish
Leaves   Alternate, compound with 5 leaflets, leaf-blade oblong to obovate, adaxial surface smooth and green, abaxial surface whitish, apex acute, margin dentate, stipules linear
Inflorescence   Panicle-like cyme
Flower   Bisexual, sepals short, triangular, hairy, petals obovate, white to pink
Fruit   Drupelets
Pollination   Entomophily
Seed dispersal   Zoochory
Habitat   Grasslands, roadsides, gardens, orchards, agri-fields, forests, hillslopes

Current status   Invasive
Impacts   Decreases native plant diversity; reduces the agricultural productivity, declines the recreational value of public parks and reserves

Native range   Africa, Europe
Global distribution   Africa, Asia-Tropical, Europe, Northern America, Southern America, Australasia

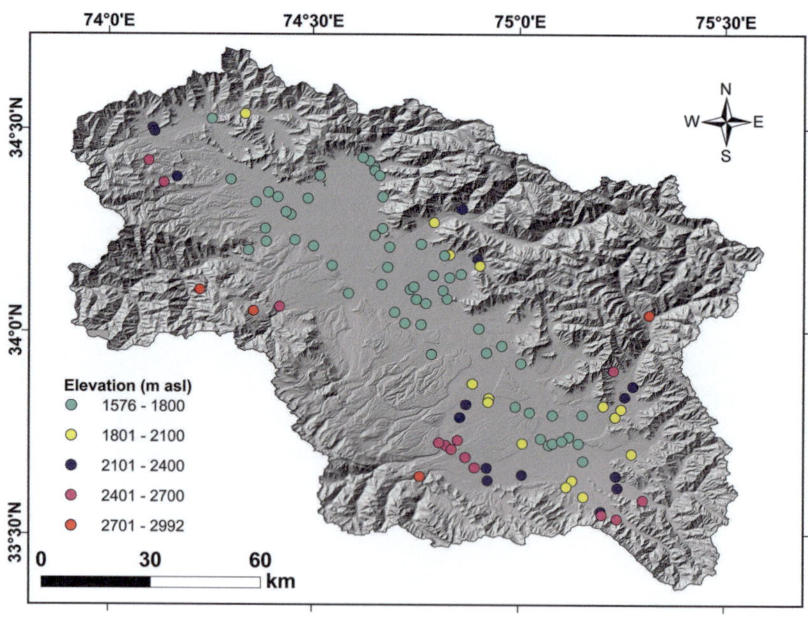

Rubus ulmifolius: (**a**) Habit, (**b**) Leaf, (**c**) Flower, (**d**) Fruit

Rudbeckia hirta L., Sp. Pl. 2: 90 (1753).

Family	Asteraceae
English names	Black-eyed susan, Yellow oxeye daisy
Local name	'Sorm-i itchal dazy'
Habit	Annual or perennial herb, 30–100 cm tall
Stem	Erect, branched, hairy
Leaves	Alternate, leaf-blade lanceolate or ovate, hairy, base attenuate, apex acute, margin entire or serrate, basal leaves petiolate, cauline leaves sessile
Inflorescence	Capitula, involucre hemispherical or ovoid
Flower	Ray florets 8–16, corolla limbs yellow to orange, base with a maroon notch, disc florets numerous, corolla brown purple
Fruit	Achene
Pollination	Entomophily
Seed dispersal	Anemochory, ornithochory
Habitat	Gardens, roadsides, orchards, forests, grasslands
Current status	Naturalised
Native range	Northern America
Global distribution	Asia-Temperate, Europe, Northern America, Southern America

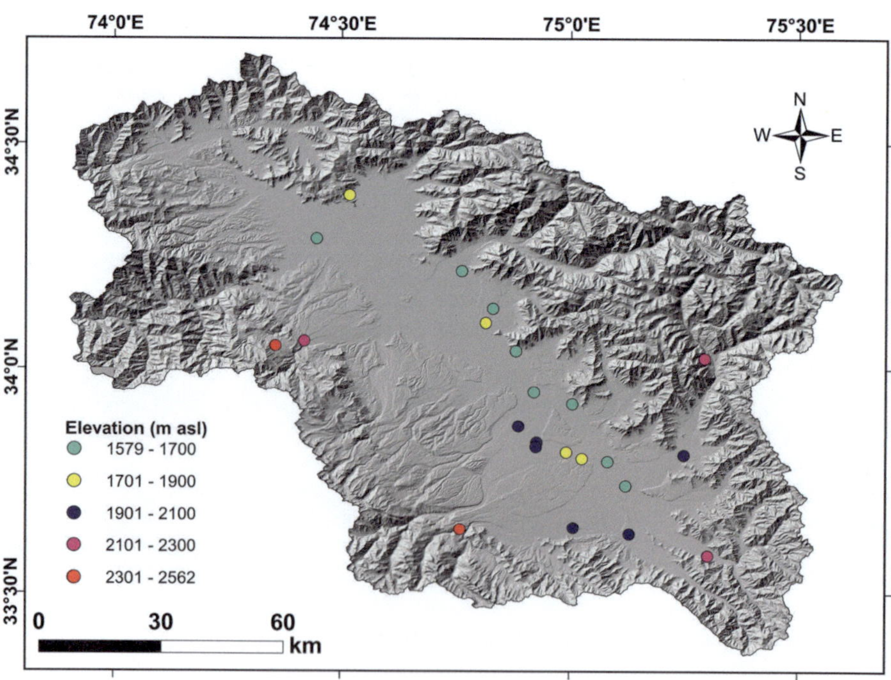

Rudbeckia hirta: (**a**) Habitat, (**b**) Habit, (**c**) Leaf, (**d**) Floral bud, (**e**) Flower (Inflorescence)

Rudbeckia laciniata L., Sp. Pl. 2: 906 (1753).

Family	Asteraceae
English names	Coneflower, Cutleaf coneflower
Local name	'Tchatich wother dazy'
Habit	Perennial herb, 100–250 cm tall
Stem	Erect, branched, glabrous
Leaves	Alternate, leaf-blade ovate to lanceolate, pinnatifid, 3–5 lobed in basal leaves, trilobed in cauline leaves, surface glabrous, base cuneate or attenuate, apex acute to acuminate, margin serrate, entire in some upper leaves
Inflorescence	Capitula, involucre hemispherical to ovoid
Flower	Ray florets 8–15, lamina ovate to lanceolate, tip bifid, disc florets numerous, yellow
Fruit	Achene
Pollination	Entomophily
Seed dispersal	Anemochory, zoochory
Habitat	Grasslands, roadsides, gardens, orchards, forests, riparian areas
Current status	Naturalised
Impacts	Decreases native plant diversity; reported to be toxic to livestock
Native range	Northern America
Global distribution	Asia-Temperate, Europe, Northern America, Australasia

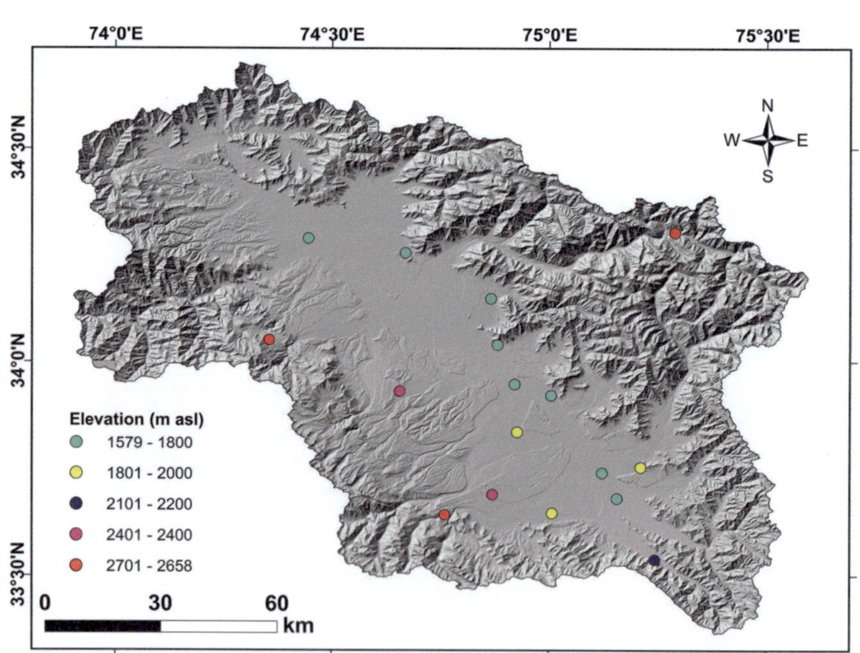

Rudbeckia laciniata: (**a**) Habitat, (**b**) Habit, (**c**) Leaf, (**d**) Flower (Inflorescence)

Saponaria officinalis L., Sp. Pl. 1: 408 (1753).

Family	Caryophyllaceae
English name	Common soapwort
Local name	'Sabni posh'
Habit	Perennial herb, 30–70 cm tall
Stem	Simple or branched above, glabrous
Leaves	Opposite, sessile, leaf-blade ovate to lanceolate, base attenuate, apex acute, margin serrulate
Inflorescence	Cyme
Flower	Pedicellate, bisexual, large, calyx green or purple, tubular, tip red, toothed, corolla of 5 petals, white or pinkish
Fruit	Capsule
Pollination	Entomophily
Seed dispersal	Autochory
Habitat	Gardens, orchards, roadsides
Current status	Naturalised
Impacts	Decreases native plant diversity; reported to be poisonous due to high saponin content
Native range	Asia-Temperate, Europe
Global distribution	Africa, Asia-Temperate, Asia-Tropical, Europe, Northern America, Southern America, Australasia

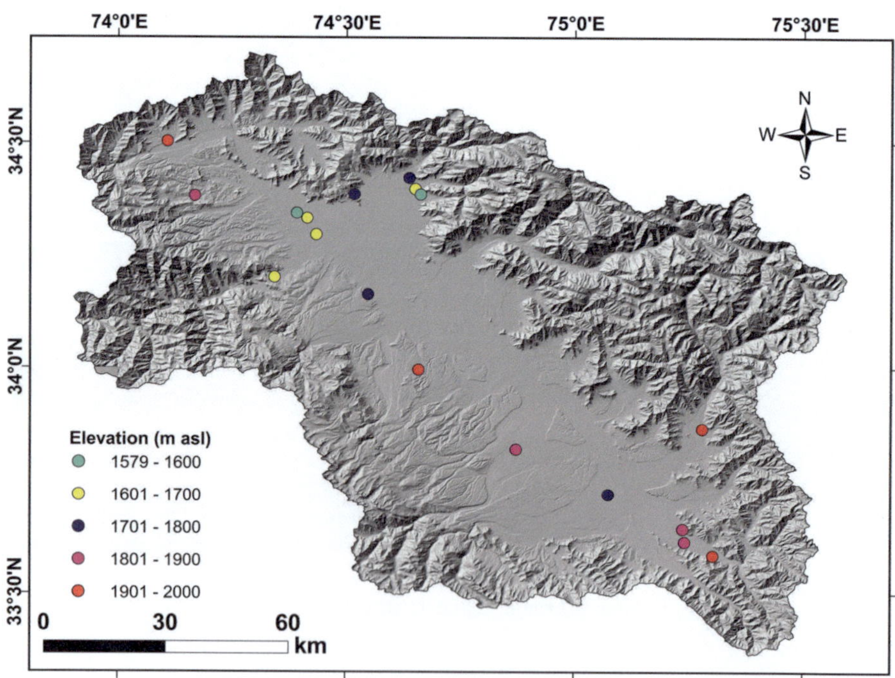

Saponaria officinalis: (**a**) Habit, (**b**, **c**) Leaf, (**d**) Inflorescence

Sedum album L., Sp. Pl. 1: 432 (1753).

Family	Crassulaceae
English name	White stonecrop
Local name	'Safeed sedum'
Habit	Perennial herb
Stem	Creeping, branched, lengthen and becomes erect at the time of flowering
Leaves	Alternate, sessile, leaf-blade linear to ovate, reddish, base spurred, apex obtuse, margin entire
Inflorescence	Paniculate cymes
Flower	Star-shaped, sepals green, connate at base, petals 5, spreading, white, rarely pinkish, filaments white, anthers red
Fruit	Follicle
Pollination	Entomophily, autogamy
Seed dispersal	Autochory
Habitat	Forests, gardens, agri-fields, grasslands, orchards
Current status	Casual
Impacts	Declines the quantity of crop yield, decreases native plant diversity, reduces the forage value of pastures
Native range	Africa, Asia-Temperate, Europe
Global distribution	Africa, Asia-Temperate, Asia-Tropical, Europe

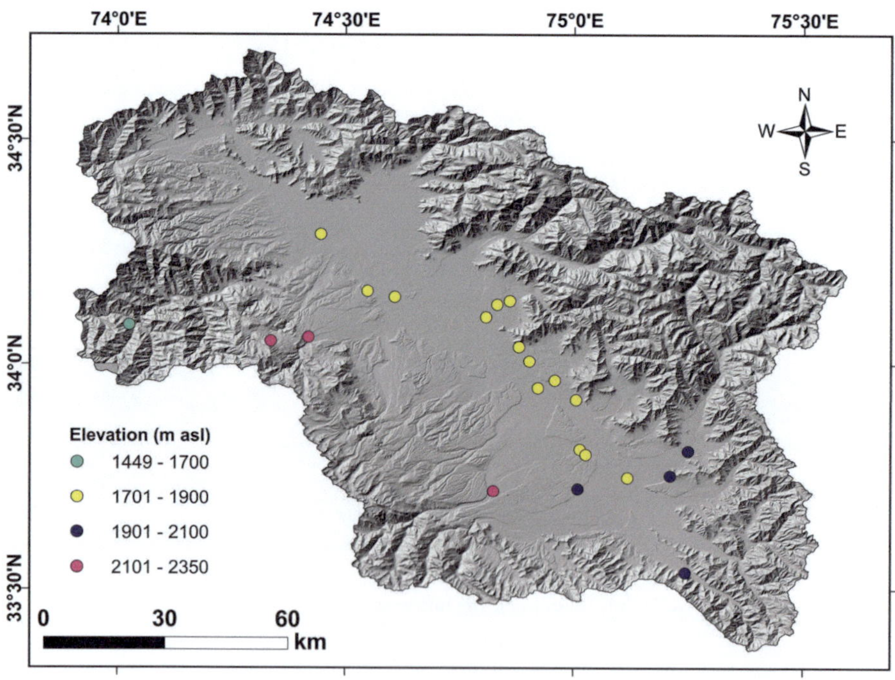

Sedum album: (**a**) Habit & Habitat, (**b**) Inflorescence, (**c**) Flower

Senecio vulgaris L., Sp. Pl. 2: 867 (1753).

Family	Asteraceae
English name	Common groundsel
Local name	'Wondi gassi'
Habit	Annual herb, 10–45 cm tall
Stem	Erect, branched, sparsely hairy
Leaves	Leaves sessile, leaf-blade oblanceolate or oblong, shallowly or deeply pinnatifid, pubescent, lateral lobes irregularly dentate, middle leaves sub-amplexicaul, apex obtuse
Inflorescence	Capitula terminal, discoid, sparsely pubescent, involucre cylindric- or urn-shaped
Flower	Ray florets absent, disc florets yellow, cylindrical, bright florets are hidden by the characteristic bract giving it the appearance of never opening flowers
Fruit	Achene
Pollination	Entomophily
Seed dispersal	Anemochory, zoochory
Habitat	Agri-fields, roadsides, gardens, grasslands, orchards, forests
Current status	Invasive
Impacts	Decreases native biodiversity, reduces the quantity and quality of crop yield, allelopathic, reported to be toxic to livestock
Native range	Africa, Asia-Temperate, Asia-Tropical, Europe
Global distribution	Africa, Asia-Temperate, Asia-Tropical, Europe, Northern America, Pacific, Southern America, Australasia

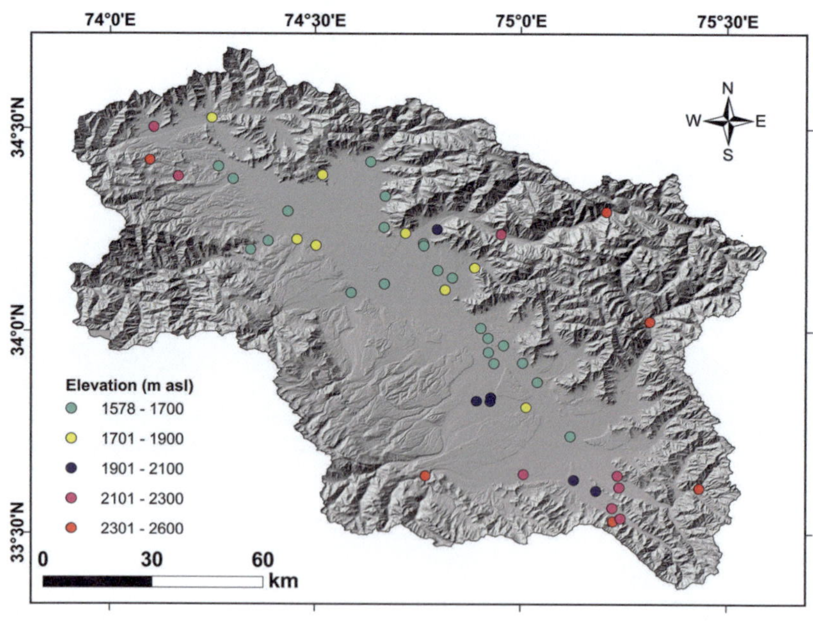

Senecio vulgaris: (**a**) Habit, (**b**) Leaf, (**c**, **d**) Inflorescence, (**e**) Fruit

Sisymbrium officinale (L.) Scop., Fl. Carniol. ed. 2. 2: 26 (1772).

Family	Brassicaceae
English name	Hedge mustard
Local name	'Manz-ni gassi'

Habit	Annual herb, upto 60 cm tall
Stem	Erect, branched, light green, hairy
Leaves	Basal leaves rosulate, petiolate, pinnatisect, cauline leaves alternate, three-lobed, margin dentate
Inflorescence	Branched raceme
Flower	Pedicellate, 4 green sepals, 4 yellow petals, style short, stamens numerous
Fruit	Siliqua
Pollination	Entomophily, autogamy
Seed dispersal	Anemochory, zoochory
Habitat	Grasslands, agri-fields, orchards, roadsides, gardens, forests

Current status	Invasive
Impacts	Reduces the quantity and quality of crop yield, decreases the native plant diversity; decreases the recreational value of natural areas, gardens and parks

Native range	Africa, Asia-Temperate, Asia-Tropical, Europe
Global distribution	Africa, Asia-Temperate, Asia-Tropical, Europe, Northern America, Pacific, Southern America, Australasia

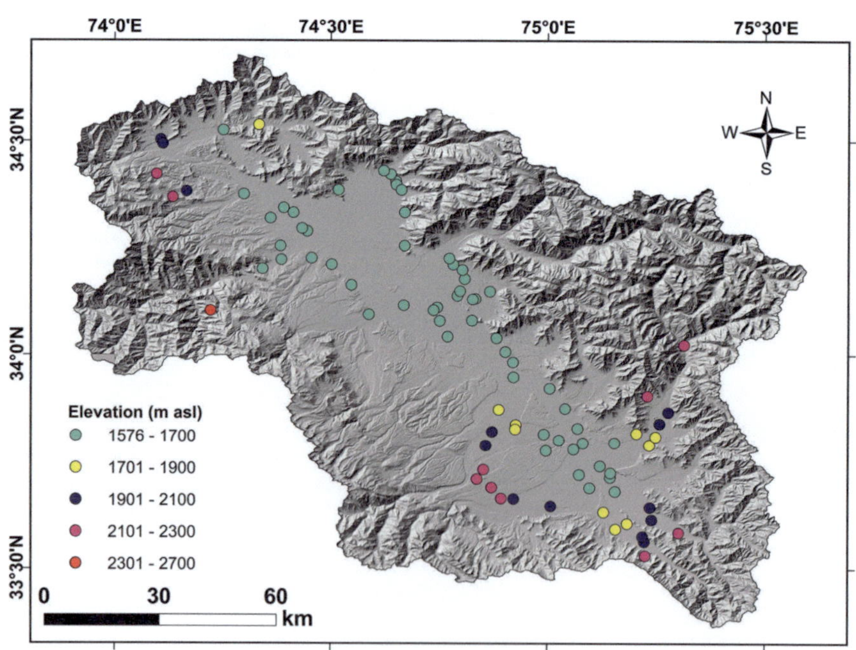

Sisymbrium officinale: (**a**) Habit, (**b**) Leaf, (**c**) Inflorescence, (**d**) Fruit

Solanum americanum Mill., Gard. Dict., ed. 8. [unpaged] Solanum no. 5 (1768).

Family	Solanaceae
English name	American black nightshade
Local name	'Kam-bai'

Habit	Annual or perennial herb, 30–150 cm tall.
Stem	Erect, branched, glabrous or sparsely pubescent
Leaves	Spiral to almost opposite, petiolate, leaf-blade ovate, base cuneate, apex acute, margin entire or dentate
Inflorescence	Sub-umbellate cyme, 3–10 flowered
Flower	Pedicellate, bisexual, calyx cup-shaped, corolla white, rarely bluish or purple
Fruit	Berry
Pollination	Entomophily
Seed dispersal	Ornithochory
Habitat	Orchards, gardens, roadsides, grasslands, agri-fields, forests

Current status	Invasive
Impacts	Decreases native plant diversity; reduces the quantity and quality of crop yield, reported to cause nausea, dry mouth, abdominal pain and vomiting in humans

Native range	Northern America, Southern America
Global distribution	Asia-Temperate, Asia-Tropical, Europe, Northern America, Southern America, Australasia

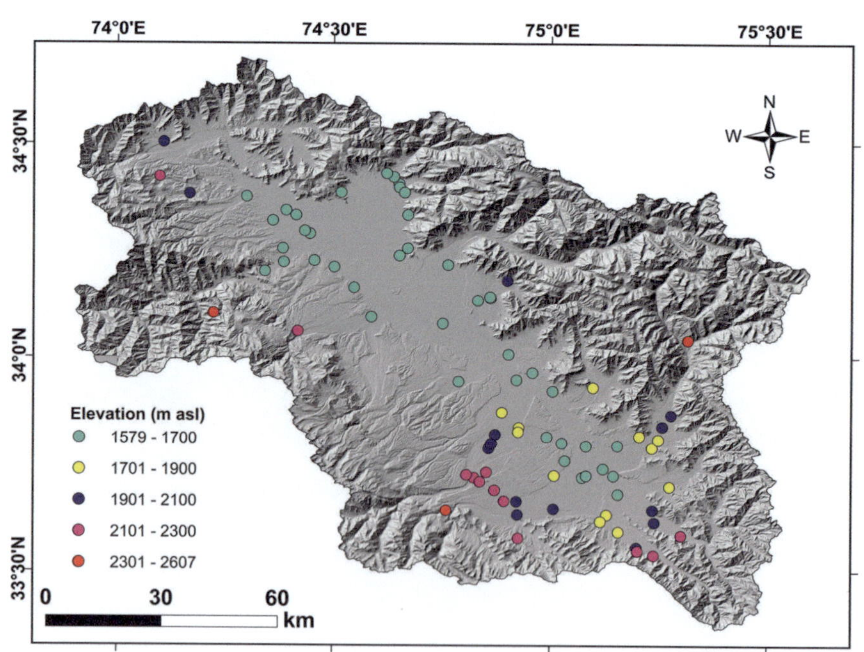

***Solanum americanum*:** (**a**) Habit, (**b**) Leaf, (**c**) Inflorescence, (**d**) Fruit

Sonchus oleraceus L., Sp. Pl. 2: 794 (1753).

Family	Asteraceae
English name	Common sow-thistle
Local name	'Dadiej'

Habit	Annual herb, upto 180 cm tall
Stem	Erect, glabrous, branched at the top, white sap oozes when broken
Leaves	Alternate, leaf-blade obovate to lanceolate, base amplexi-caul, apex round, margin weakly spiny
Inflorescence	Capitula, involucre urn-shaped, bracts green, lance-shaped
Flower	Ligulate, corolla yellow
Fruit	Cypsela
Pollination	Entomophily, anemophily
Seed dispersal	Anemochory, zoochory
Habitat	Agri-fields, roadsides, gardens, grasslands, orchards, forests

Current status	Invasive
Impacts	Decreases native plant diversity; reduces the quantity and quality of crop yield, reported to act as an alternate host to many insect pests and viral diseases

Native range	Africa, Asia-Temperate, Asia-Tropical, Europe
Global distribution	Africa, Asia-Temperate, Asia-Tropical, Northern America, Pacific, Southern America, Australasia

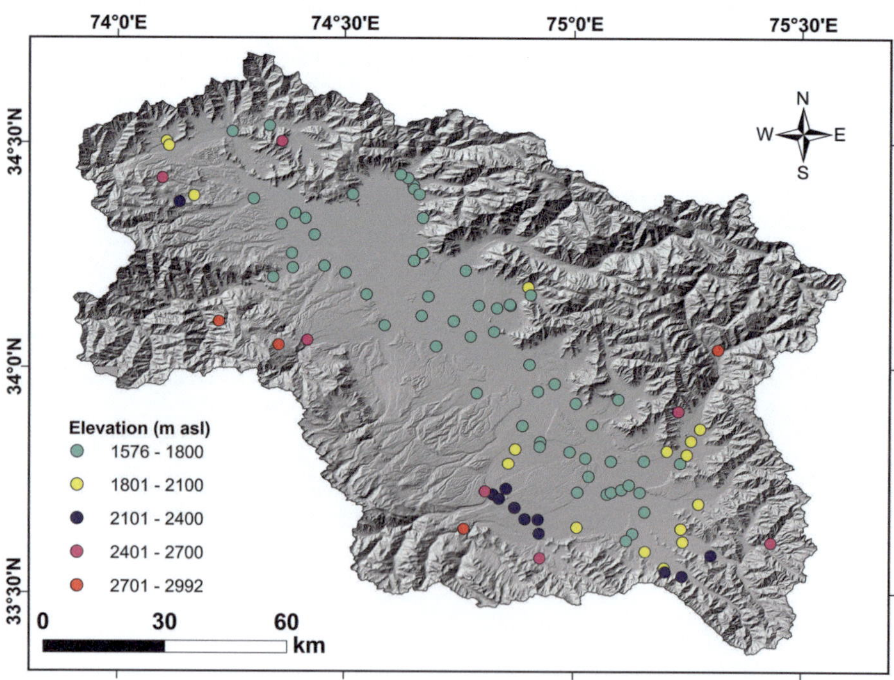

Sonchus oleraceus: (**a**) Habit & Habitat, (**b**) Leaf, (**c, d**) Flower (Inflorescence)

Spartium junceum L., Sp. Pl. 2: 708 (1753).

Family	Fabaceae
English name	Spanish broom
Local name	'Panjeab kani kul'
Habit	Perennial shrub, 300–400 cm tall
Stem	Erect, branched, cylindrical, green when young
Leaves	Either absent or very small, oval-shaped
Inflorescence	Raceme
Flower	Large, pea flower like, yellow
Fruit	Legume (Pod)
Pollination	Entomophily
Seed dispersal	Anemochory, zoochory
Habitat	Roadsides, gardens, hill slopes, forests, riparian areas
Current status	Naturalised
Impacts	Reduces native plant diversity; decreases fodder potential of pasturelands; increases risk and intensity of fires in forests and shrublands
Native range	Europe
Global distribution	Africa, Asia-Temperate, Asia-Tropical, Europe, Southern America, Australasia

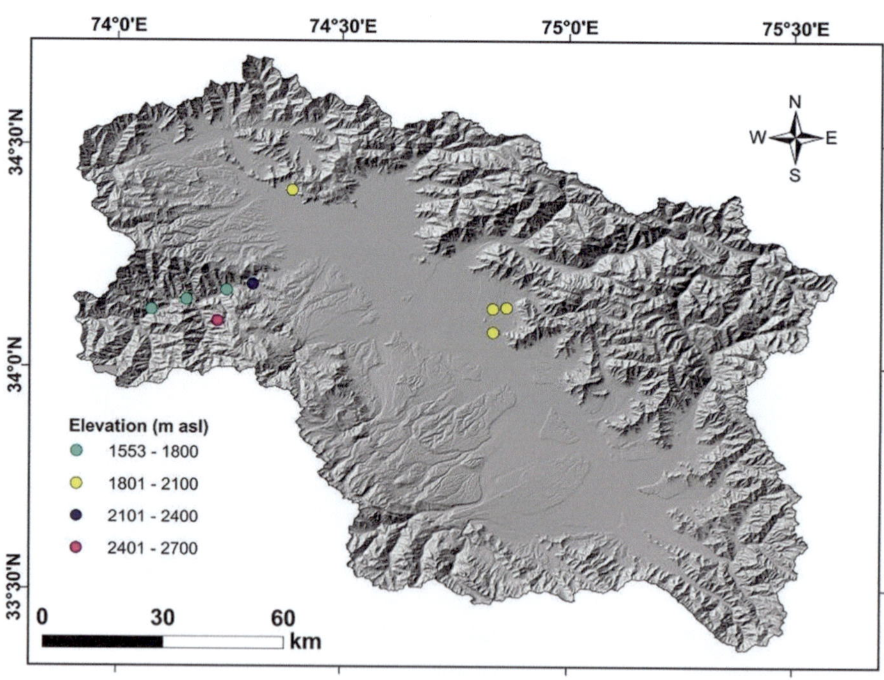

Spartium junceum: (**a**) Habitat, (**b**) Inflorescence, (**c**) Flower, (**d**) Fruit

Symphyotrichum subulatum (Michx.) G.L.Nesom, Phytologia 77(3): 293 (1995).

Family	Asteraceae
English name	Annual saltmarsh aster
Local name	'Damboel gassi'
Habit	Annual herb, 16–200 cm tall
Stem	Stem erect, branched, glabrous, green, becomes purplish on maturity
Leaves	Leaves two types—basal leaves petiolate, lanceolate to ovate, wither at anthesis, cauline leaves sessile, linear to lanceolate, becoming smaller in size upwards, adaxial and abaxial leaf surfaces glabrous, base attenuate, margin serrulate to entire, apex acute
Inflorescence	Inflorescence capitula in paniculiform synflorescences, involucre cylindrical, phyllaries 3–4 seriate, lanceolate, clearly unequal, margin entire, tip acute and prominently purplish
Flower	Ray florets 2–3 seriate, 34–36 in number, lamina bluish or pinkish-white, coiling backwards, stigma bifid, bristles whitish, disc florets 12–14 in number, yellow, lobes erect and triangular, stigma feathery and spoon-shaped
Fruit	Achene
Pollination	Entomophily
Seed dispersal	Anemochory
Habitat	Agri-fields, roadsides, gardens, grasslands, orchards, riparian areas, forests
Current status	Invasive
Impacts	Decreases native plant diversity; reduces the quantity and quality of crop yield, reduces the forage value of pastures
Native range	Northern America, Southern America
Global distribution	Africa, Asia-Temperate, Europe, Northern America, Pacific, Southern America, Australasia

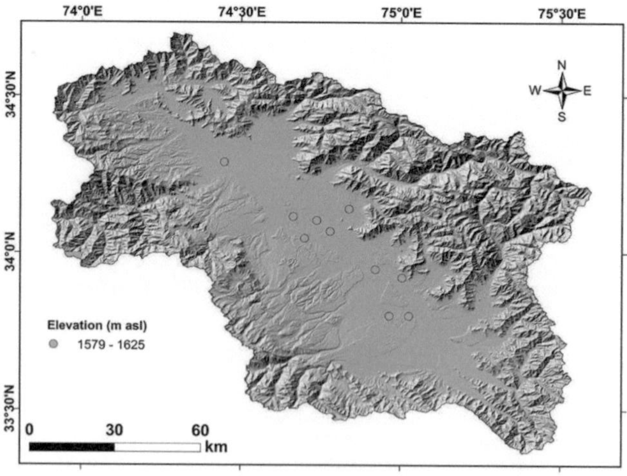

Symphyotrichum subulatum: (**a, b**) Habit & Habitat, (**c, d**) Flower (Inflorescence), (**e, f**) Phyllaries, (**g**) Fruit

Tagetes erecta L., Sp. Pl. 2: 887 (1753).

Family	Asteraceae
English name	Marigold
Local name	'Batipoosh'
Habit	Annual herb, 30–150 cm tall
Stem	Erect, branched, glabrous
Leaves	Lower leaves opposite, upper leaves alternate, pinnatisect, lobes linear to lanceolate or lanceolate, surface glabrous, margin serrate
Inflorescence	Capitula, phyllaries 5–8
Flower	Ray florets usually 5–8 or upto 100 in some cases, lamina yellow, orange or white, disc florets more than 100
Fruit	Achene
Pollination	Entomophily
Seed dispersal	Autochory, anemochory, zoochory
Habitat	Gardens, roadsides, grasslands
Current status	Casual
Impacts	Allelopathic, decreases the native plant diversity; reported to cause dermatitis, burning pain, skin redness and blisters in humans
Native range	Northern America, Southern America
Global distribution	Africa, Asia-Tropical, Europe, Northern America, Pacific, Southern America, Australasia

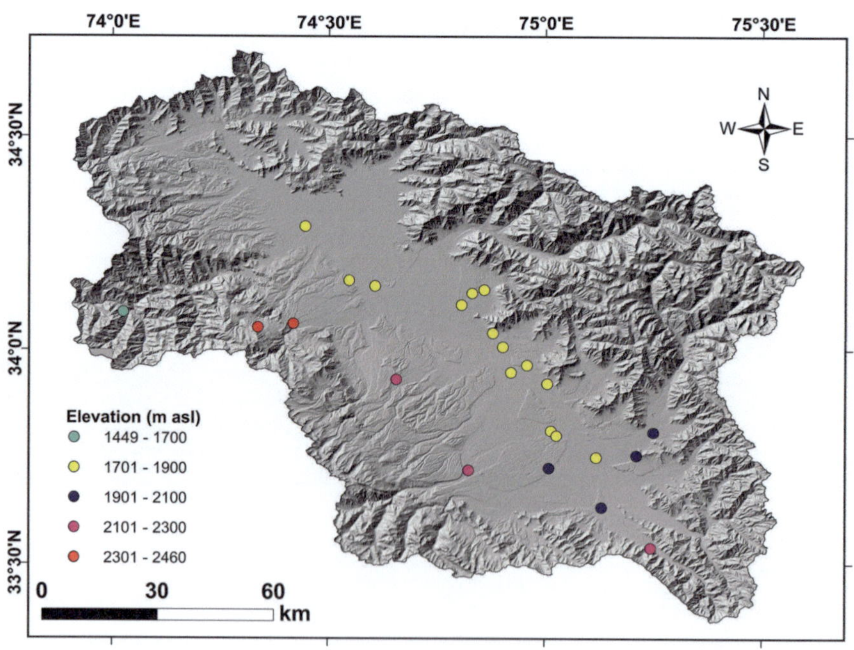

***Tagetes erecta*:** (**a**) Habit, (**b**) Leaf, (**c**) Floral bud, (**d**) Flower (Inflorescence)

Tagetes minuta L., Sp. Pl. 2: 887 (1753).

Family	Asteraceae
English name	Wild marigold
Local name	'Jangle jaafiry'
Habit	Annual herb, upto 200 cm tall
Stem	Erect, much branched, ribbed, glabrous
Leaves	Opposite, alternate in upper part, petiolate, pinnatisect, lobes linear to lanceolate, margin serrate
Inflorescence	Capitula in corymbiform clusters, involucre cylindric
Flower	Ray florets 3–4, creamy or pale yellow, disc florets 4–7, yellow
Fruit	Achene
Pollination	Entomophily
Seed dispersal	Anemochory, zoochory
Habitat	Agri-fields, roadsides, gardens, orchards, forests
Current status	Invasive
Impacts	Decreases native plant diversity, allelopathic, decreases quality and quantity of crop yield, especially maize and beans, reported to cause skin irritation to agricultural workers, gives irritable flavour to milk if consumed by cattle
Native range	Southern America
Global distribution	Africa, Asia-Temperate, Asia-Tropical, Europe, Northern America, Pacific, Southern America, Australasia

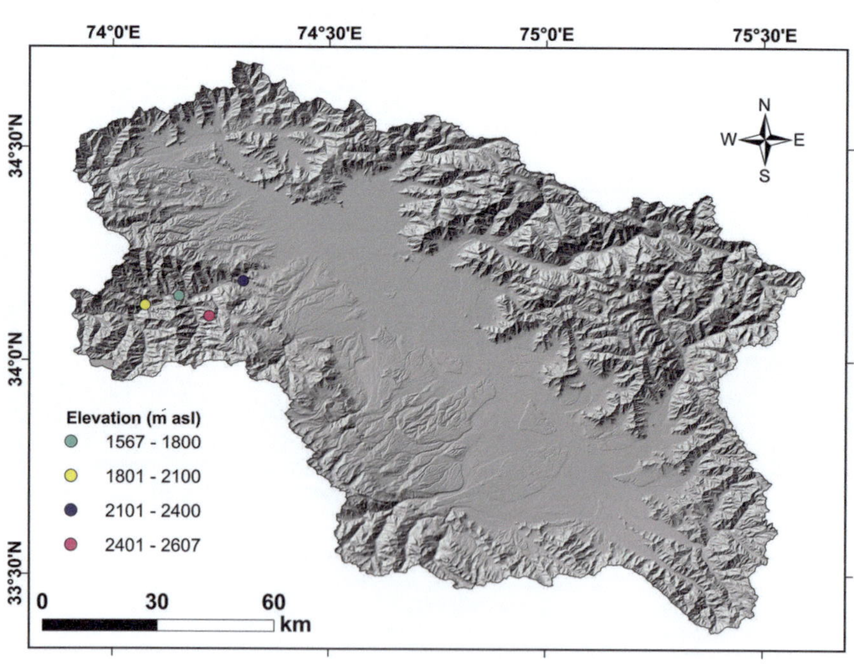

Tagetes minuta: (**a**) Habit & Habitat, (**b**) Leaf, (**c**, **d**) Inflorescence

Tagetes tenuifolia Cav., Icon. [Cavanilles] 2(2): 54, t. 169 (1793).

Family	Asteraceae
English names	Signet marigold, Golden marigold
Local name	'Son-i jaafiry'
Habit	Annual herb, 60–80 cm tall
Stem	Erect, branched, glabrous or sparsely hairy
Leaves	Alternate, shortly petiolate or sessile, pinnately compound, leaf segments lanceolate, both surfaces glabrous, glandular dots on lower surface, margin serrate
Inflorescence	Capitula, involucre cylindrical
Flower	Sepals green, fused upto the top, tip of sepals acute, ray florets 5–8, velvety red, disc florets numerous
Fruit	Achene
Pollination	Entomophily
Seed dispersal	Autochory, anemochory, zoochory
Habitat	Gardens, roadsides, grasslands, forests
Current status	Casual
Native range	Northern America, Southern America
Global distribution	Asia-Temperate, Northern America, Southern America

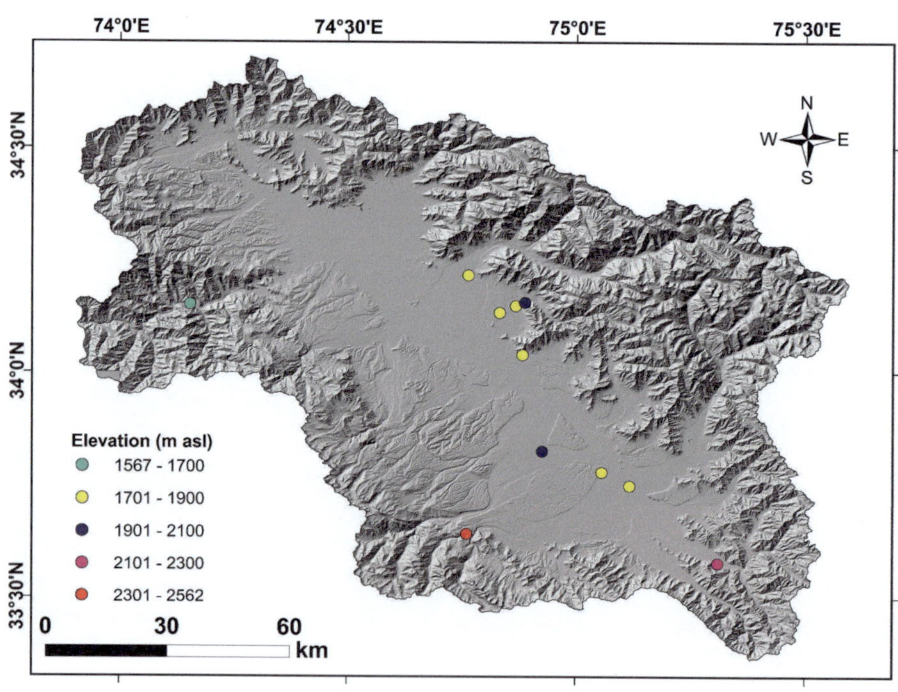

Tagetes tenuifolia: (**a**) Habit & Habitat, (**b**) Leaf, (**c**) Floral bud, (**d**) Flower (Inflorescence)

Tanacetum vulgare L., Sp. Pl. 2: 844 (1753).

Family	Asteraceae
English name	Common tansy
Local name	'Aam tansy'

Habit	Perennial herb, 40–150 cm tall
Stem	Erect, single or clustered, branched, glabrous
Leaves	Lower stem leaves petiolate, leaf-blade elliptic to ovate, pinnatisect, ultimate segments ovate, triangular or narrowly elliptic, upper stem leaves similar but sessile
Inflorescence	Synflorescence in the form of panicle, capitula heterogamous, involucre campanulate, phyllaries 3-seriate
Flower	All florets yellow, outer florets female, inner disc florets bisexual
Fruit	Achene
Pollination	Entomophily
Seed dispersal	Ornithochory, zoochory
Habitat	Gardens, orchards

Current status	Casual
Impacts	Reduces native plant diversity in habitats where it becomes dominant; reported to be toxic to livestock

Native range	Asia-Temperate, Europe
Global distribution	Asia-Temperate, Europe, Northern America, Southern America, Australasia

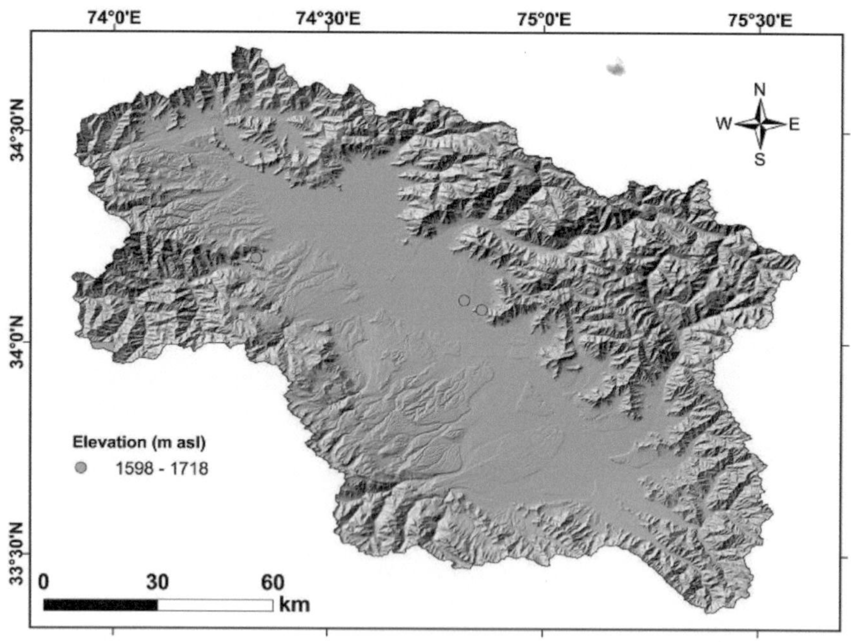

Tanacetum vulgare: (**a**) Habit & Habitat, (**b**) Leaf, (**c**) Flower (Inflorescence)

Trifolium dubium Sibth., Fl. Oxon. 231. (1794).

Family	Fabaceae
English names	Lesser-hop trefoil, Suckling clover
Local name	'Parim chati bati'
Habit	Annual herb
Stem	Erect or procumbent, branched, pubescent
Leaves	Alternate, petiolate, trifoliate, leaflets obovate or obcordate, margin minutely serrate, Stipules ovate
Inflorescence	Raceme, hemispherical in shape
Flower	Pedicellate, pea flower like, calyx short, glabrous, corolla long, yellow or yellowish
Fruit	Pod
Pollination	Entomophily
Seed dispersal	Autochory
Habitat	Agri-fields, roadsides, gardens, orchards, forests
Current status	Invasive
Impacts	Decreases native plant diversity; reduces the crop yield, reported to alter the nitrogen level of soil
Native range	Asia-Tropical, Northern America
Global distribution	Africa, Asia-Temperate, Asia-Tropical, Europe and Northern America

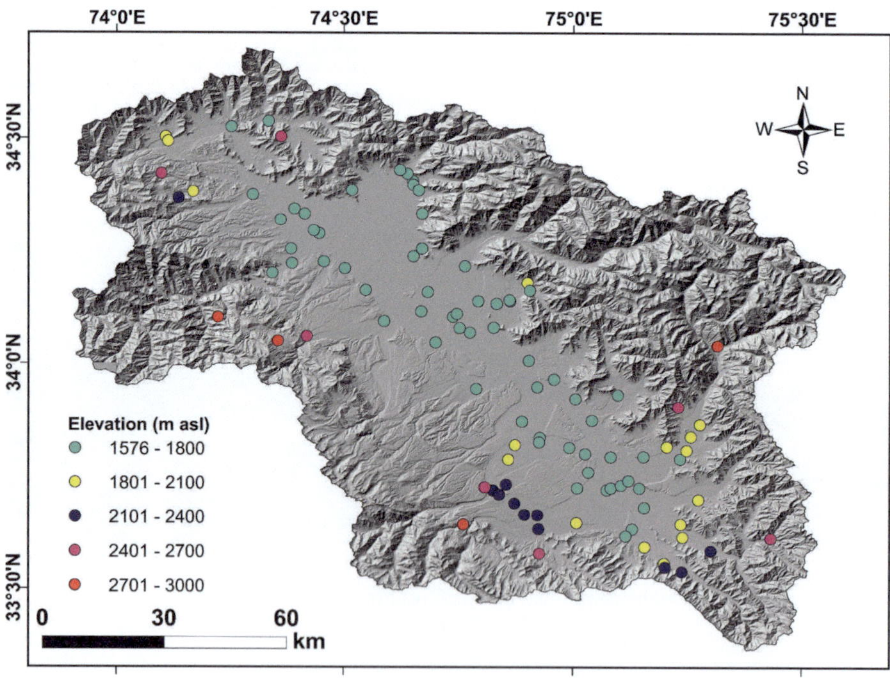

Trifolium dubium: (**a**) Habit, (**b**) Leaf, (**c**) Flower

Veronica hederifolia L., Sp. Pl. 1: 13 (1753).

Family	Plantaginaceae
English name	Ivy-leaf speedwell
Local name	'Chari gassi'
Habit	Annual herb, upto 50 cm tall
Stem	Prostrate, branched, pubescent
Leaves	Lower stem leaves opposite, petiolate, leaf-blade ovate, glabrous, upper stem leaves petiolate, blade 3–5 lobed, pubescent
Inflorescence	Solitary, axillary
Flower	Pedicellate, calyx four-lobed, lobes unequal, ovate, margin ciliate, corolla blue, petals obovate, glabrous
Fruit	Capsule
Pollination	Entomophily
Seed dispersal	Autochory, myrmecochory
Habitat	Gardens, roadsides, grasslands, orchards, agri-fields, forests
Current status	Naturalised
Impacts	Allelopathic, decreases the quality and quantity of crop yield; decreases native plant diversity
Native range	Africa, Asia-Temperate, Europe
Global distribution	Africa, Asia-Temperate, Europe, Northern America

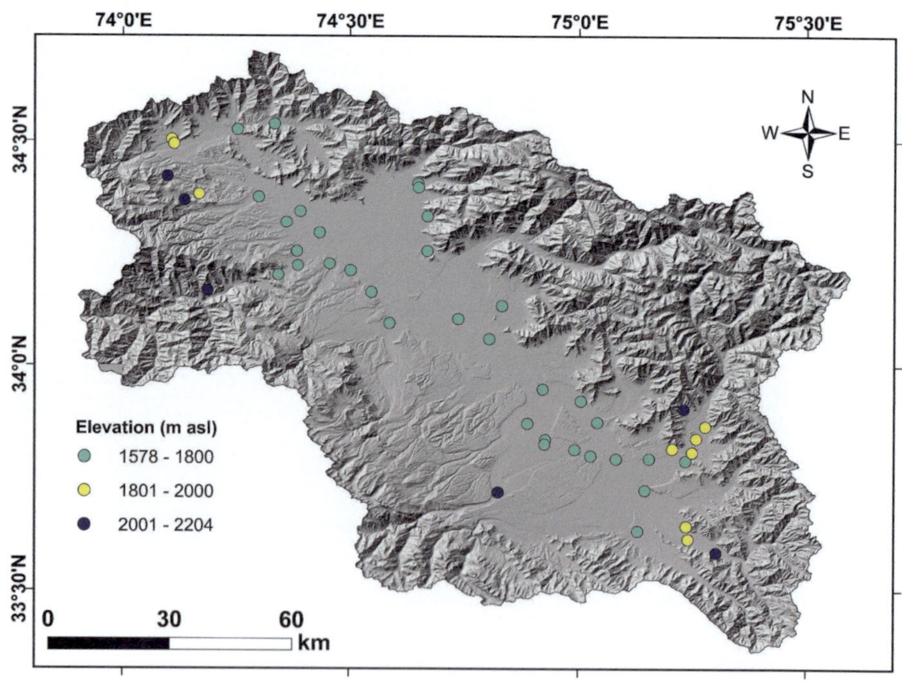

Veronica hederifolia: (**a**) Habit, (**b**) Basal leaf, (**c**) Cauline leaf, (**d**) Calyx, (**e**) Flower

Veronica persica Poir., Encycl. [J. Lamarck & al.] 8: 542 (1808).

Family	Plantaginaceae
English names	Persian speedwell, Common field speedwell
Local name	'Zovi gassi'
Habit	Annual herb
Stem	Prostrate to ascending, 10–70 cm tall, pubescent
Leaves	Opposite, shortly petiolate, leaf-blade ovate to suborbicular, margin crenate—serrate
Inflorescence	Solitary, axillary, bracts alternate, leaf like, petiolate
Flower	Pedicellate, calyx four-lobed, ovate-lanceolate, sparsely pubescent, corolla sky blue, petals ovate to sub-orbicular, upper surface with stripes, centre white
Fruit	Capsule
Pollination	Entomophily
Seed dispersal	Ornithochory
Habitat	Agri-fields, roadsides, gardens, orchards, forests
Current status	Invasive
Impacts	Decreases native plant diversity; allelopathic, decreases quality and quantity of crop yield
Native range	Southwest Asia
Global distribution	Asia-Temperate, Asia-Tropical, Europe, Northern America, Southern America, Australasia

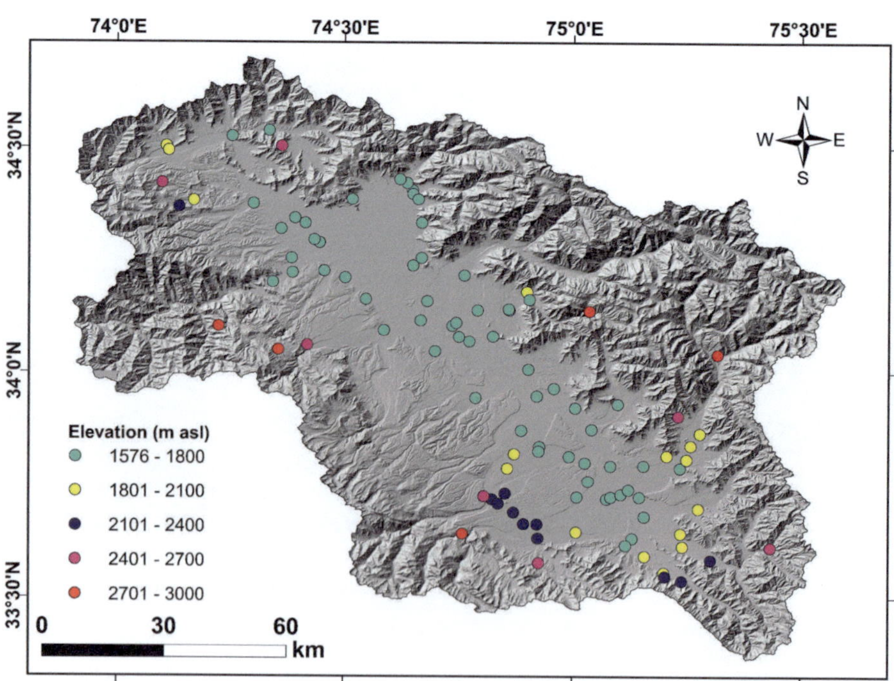

Veronica persica: (**a**) Habitat, (**b**) Habit, (**c**) Inflorescence, (**d**) Flower

Veronica polita subsp. ***lilacina*** (T.Yamaz.) T.Yamaz., Fl. Jap. (Iwatsuki et al., eds.)
3a: 354 (1993).

Family	Plantaginaceae
English name	Grey field speedwell
Local name	'Soer zovi gassi'
Habit	Annual herb, upto 15 cm long
Stem	Stem prostrate to ascending
Leaves	Lower leaves petiolate, opposite, upper leaves near the terminal inflorescence sessile and alternate, leaf blade 5–7 lobed, ovate to orbicular, sparsely pubescent to glabrous on the adaxial surface and densely pubescent on abaxial surface.
Inflorescence	Solitary terminal and axillary
Flower	Pedicellate, sepals herbaceous, ovate with overlapping base, petals whitish-pink, with 5–10 rose-purple lines towards the inner surface
Fruit	Capsule
Pollination	Entomophily
Seed dispersal	Anemochory
Habitat	Gardens, roadsides, orchards, agri-fields, forests
Current status	Naturalised
Impacts	Decreases native plant diversity, allelopathic, decreases quality and quantity of crop yield
Native range	Japan, Korea
Global distribution	Asia-Temperate

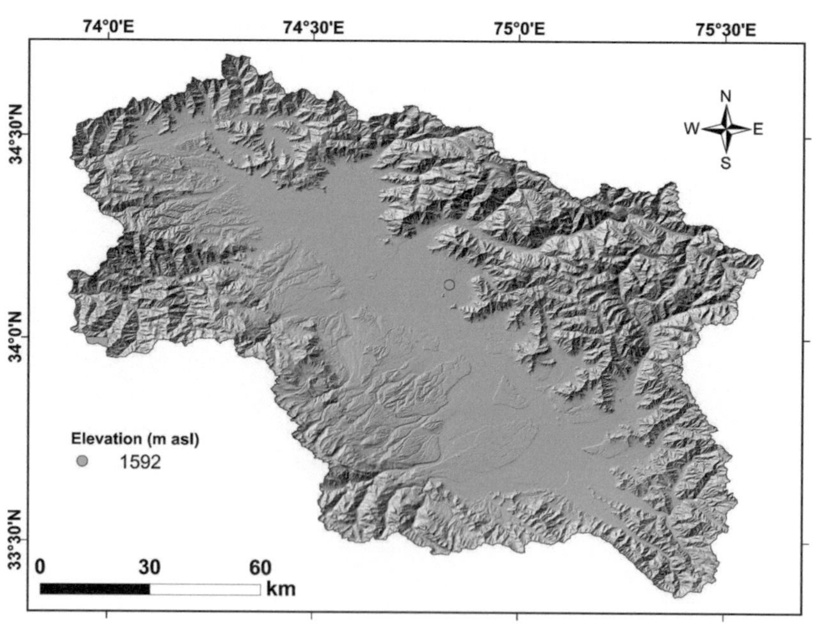

***Veronica polita* subsp. *lilacina*: (a)** Habit & Habitat, **(b)** Flower, **(c)** Calyx, **(d)** Fruit, **(e, f)** Seeds

Vinca major L., Sp. Pl. 1: 209 (1753).

Family	Apocynaceae
English name	Large-leaf periwinkle
Local name	'Boed vanca vail'
Habit	Perennial herb, sprawling
Stem	Prostrate, rooting at nodes
Leaves	Opposite, petiolate, leaf-blade ovate, large, leathery, margin entire
Inflorescence	Solitary
Flower	Blue to purple in colour, calyx campanulate, corolla enlarged above the middle, stigma capitate
Fruit	Follicle
Pollination	Entomophily
Seed dispersal	Zoochory
Habitat	Gardens, roadsides, agri-fields, orchards, forests
Current status	Naturalised
Impacts	Decreases native plant diversity; reduces the crop yield
Native range	Asia-Temperate, Europe
Global distribution	Africa, Asia-Temperate, Asia-Tropical, Europe, Northern America, Pacific, Southern America, Australasia

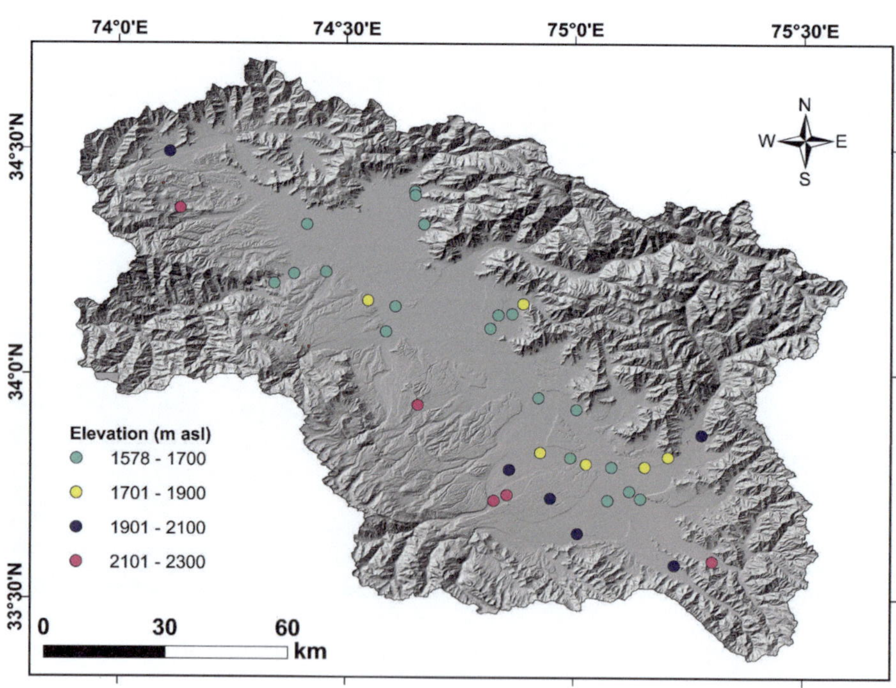

Vinca major: (**a**) Habit, (**b**) Leaf, (**c**) Floral bud, (**d, e**) Flower

Vinca minor L., Sp. Pl. 1: 209 (1753).

Family	Apocynaceae
English name	Lesser periwinkle
Local name	'Loket vanca vail'
Habit	Perennial herb, creeping
Stem	Prostrate, glabrous, rooting at nodes.
Leaves	Opposite, petiolate, leaf-blade ovate or elliptic, leathery, glabrous, margin entire
Inflorescence	Solitary, axillary
Flower	Calyx campanulate, sepals long, slender, corolla lilac to blue, base tubular, lobes obliquely truncate
Fruit	Pair of follicles
Pollination	Entomophily
Seed dispersal	Zoochory
Habitat	Gardens, roadsides, agri-fields, orchards, forests
Current status	Casual
Impacts	Decreases native plant diversity; reduces the crop yield
Nativity	Asia-Temperate, Europe
Global distribution	Asia-Temperate, Europe, Northern America, Australasia

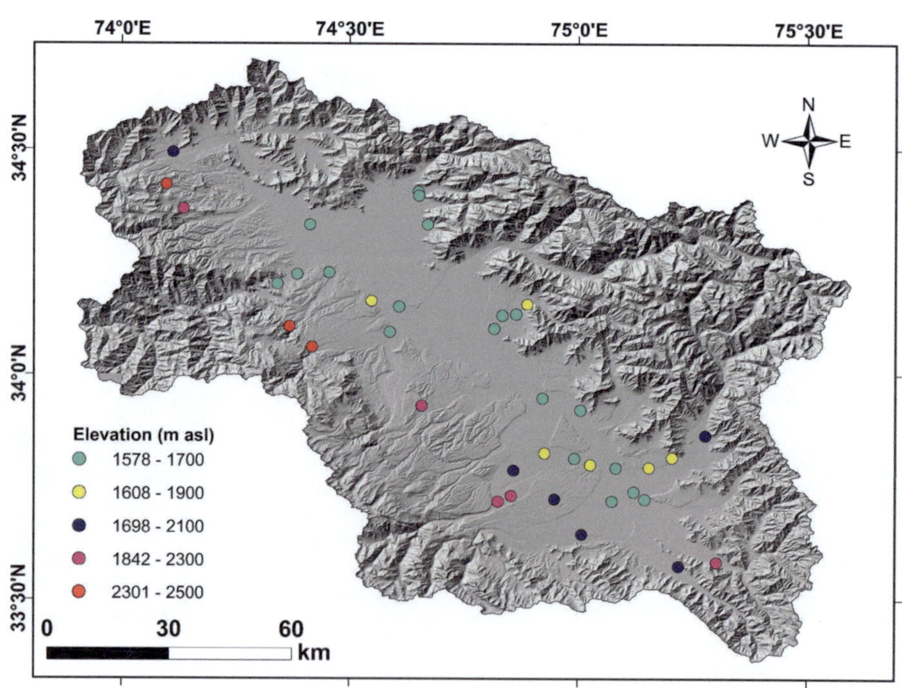

Vinca minor: (**a**) Habitat, (**b**) Habit, (**c**) Leaf, (**d**) Flower

Xanthium spinosum L., Sp. Pl. 2: 987 (1753).

Family	Asteraceae
English names	Spiny cocklebur, Prickly burweed
Local name	'Kiend dar tcheer-i kul'

Habit	Annual herb, 100–250 cm tall
Stem	Erect, lined at intervals with long, sharp, yellowish spines, which may exceed 3 cm in length and may divide into two or three separate spines
Leaves	Alternate or spiral, petiolate, leaf-blade divided into linear or lance-shaped lobes, the middle one much longer than the others, adaxially dark green or greyish, abaxially white
Inflorescence	Separate male and female flower heads, female heads develop into burs
Flower	Yellowish in colour
Fruit	Bur
Pollination	Entomophily
Seed dispersal	Zoochory
Habitat	Grasslands, agri-fields, roadsides, gardens, hillslopes, orchards, forests

Current status	Invasive
Impacts	Reduces the quantity and quality of crop yield, declines the forage value of pastures, reported to be toxic to livestock, acts as a host to many fungal diseases; prickles cause injury to humans and animals by clinging to clothes, hair and fur; decreases the native plant diversity

Native range	Southern America
Distribution	Africa, Asia-Temperate, Asia-Tropical, Europe, Northern America, Southern America, Australasia

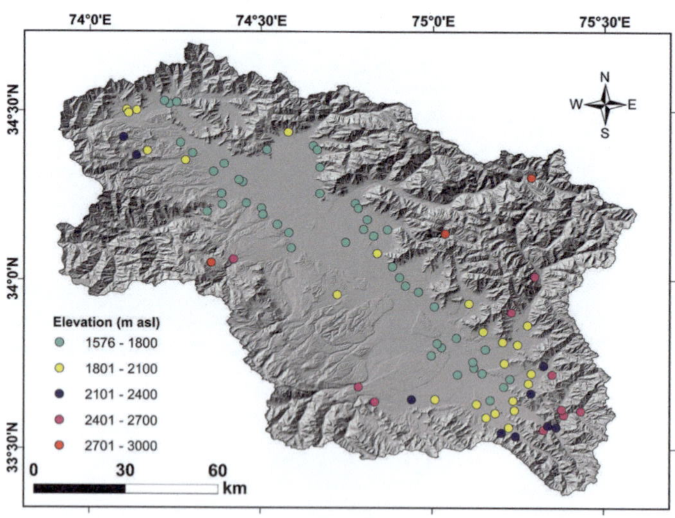

Xanthium spinosum: (**a**) Habitat, (**b**) Leaf, (**c**) Inflorescence, (**d**) Fruit

Xanthium strumarium L., Sp. Pl. 2: 987 (1753).

Family	Asteraceae
English name	Common cocklebur
Local name	'Tcheer-i kul'
Habit	Annual herb, 100–150 cm tall
Stem	Short, hairy, maroon at maturity
Leaves	Alternate, petiolate, leaf-blade ovate-deltoid, both surfaces scabrous, margin irregularly dentate
Inflorescence	Capitula in cymes, involucre connate
Flower	Corolla tubular, white or green, covered with hooked bristles
Fruit	Bur
Pollination	Entomophily
Seed dispersal	Anemochory, zoochory
Habitat	Agri-fields, grasslands, roadsides, gardens, hillslopes, orchards, forests
Current status	Invasive
Impacts	Decreases native plant diversity; reduces the quantity and quality of crop yield, declines the forage value of pastures, reported to be toxic to livestock, prickles cause injury to humans
Native range	Africa, Asia-Temperate, Asia-Tropical, Europe, Northern America, Southern America
Global distribution	Africa, Asia-Temperate, Asia-Tropical, Europe, Australasia, Northern America, Pacific, Southern America

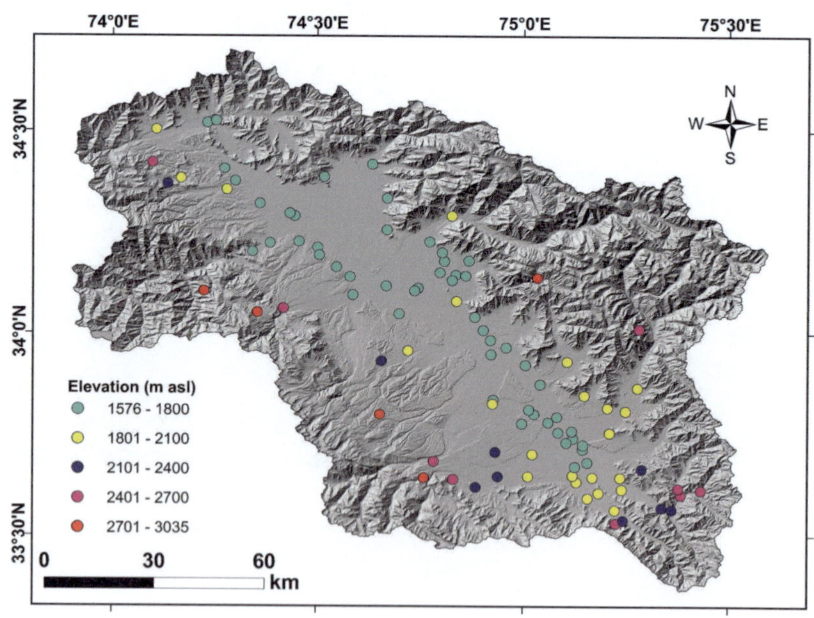

Xanthium strumarium: (**a**, **b**) Habit & Habitat, (**c**) Leaf, (**d**) Inflorescence, (**e**) Fruit

Youngia japonica (L.) DC., Prodr. [A.P. de Candolle] 7(1): 194 (1838).

Family	Asteraceae
English name	Oriental hawkweed
Local name	'Ganti katch'

Habit	Annual or biennial herb, 20–120 cm tall
Stem	Erect, branched, hairy
Leaves	Basal leaves rosette, leaf-blade lyrate or oblanceolate, margin dentate, cauline leaves few, similar to basal leaves or reduced to bracts
Inflorescence	Capitula, corymb or panicle like, involucre cylindric, phyllaries pubescent
Flower	10–20 ray florets, corolla yellow, tip dentate, disc florets absent
Fruit	Achene
Pollination	Entomophily
Seed dispersal	Anemochory, zoochory
Habitat	Grasslands, roadsides, gardens, orchards, agri-fields, forests

Current status	Invasive
Impacts	Decreases native plant diversity; reduces the quantity and quality of crop yield

Native range	Asia-Temperate, Asia-Tropical, Australasia
Global distribution	Africa, Asia-Temperate, Asia-Tropical, Northern America, Pacific, Southern America, Australasia

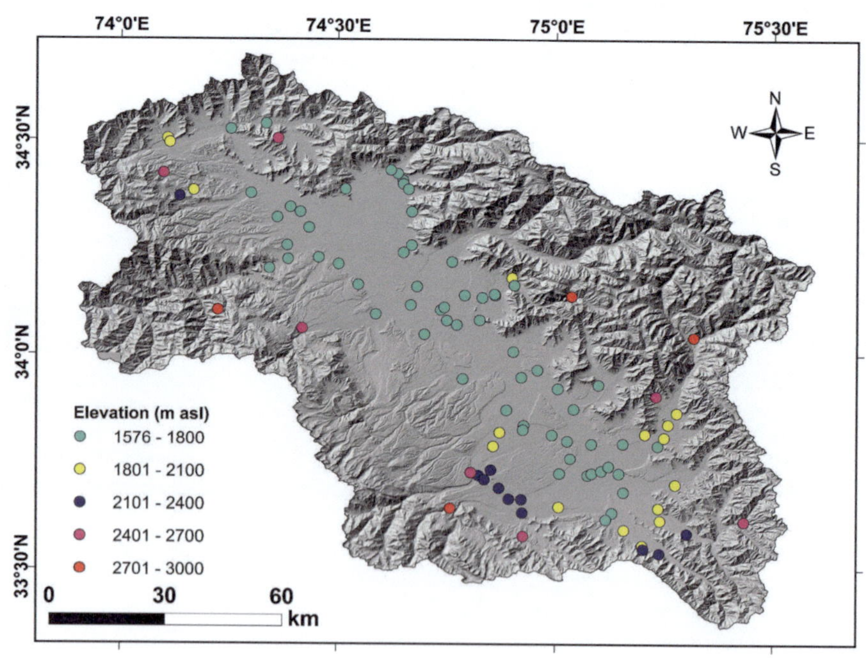

Youngia japonica: (**a**) Habit & Habitat, (**b**) Leaf, (**c**, **d**) Flower (Inflorescence)

Zinnia angustifolia Kunth, Nov. Gen. Sp. [H.B.K] 4(17): 197 (ed. fol.) (1818).

Family	Asteraceae
English name	Narrow-leaf zinnia
Local name	'Zeawul-pan jaafiry'

Habit	Annual or perennial herb, upto 60 cm tall
Stem	Erect, branched, hairy
Leaves	Opposite, sessile, leaf-blade linear to lanceolate, scabrous, base attenuate, tip acute, margin entire
Inflorescence	Capitula, involucre hemispherical
Flower	Ray florets bright orange, yellow or white, disc florets yellow
Fruit	Achene
Pollination	Entomophily
Seed dispersal	Autochory, anemochory
Habitat	Gardens, roadsides, orchards, agri-fields

Current status	Casual
Native range	Northern America
Global distribution	Northern America, Southern America

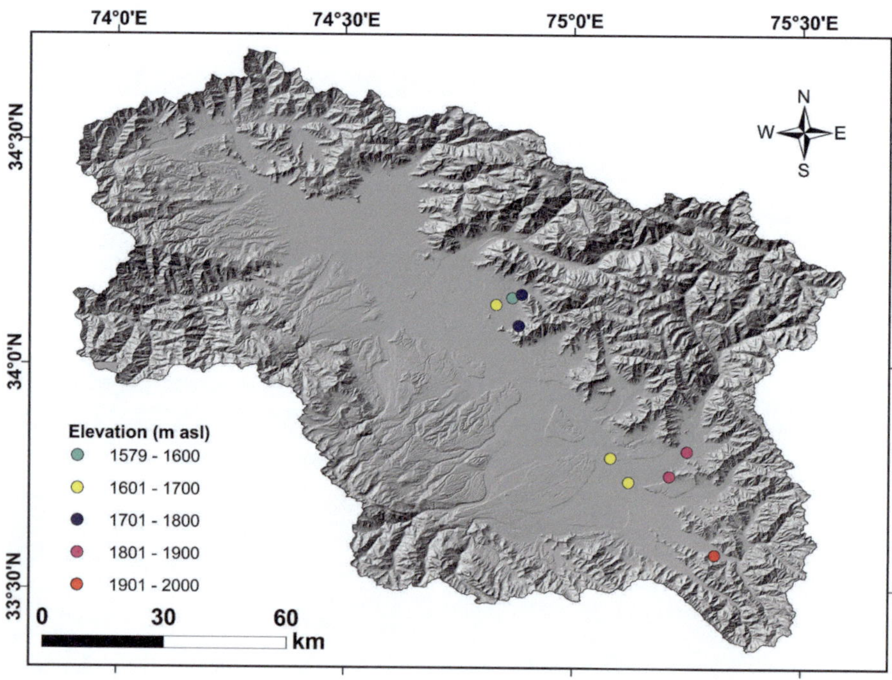

Zinnia angustifolia: (**a**) Habit, (**b**) Leaf, (**c**, **d**, **e**) Flower (Inflorescence)

Index

A
Abutilon theophrasti 24–25
Aesculus indica 26–27
Ageratum conyzoides 28–29
Ailanthus altissima 30–31
Alcea rosea 32–33
Alternanthera philoxeroides 34–35
Amaranthus
    A. caudatus 36–37
    A. spinosus 38–39
    A. viridis 40–41
Anethum graveolens 42–43
Anthemis cotula 44–45
Arctium lappa 46–47
Artemisia annua 48–49
Asparagus officinalis 50–51

B
Bellis perennis 52–53
Bidens bipinnata 54–55
Bromus catharticus 56–57
Buddleja davidii 58–59

C
Calendula officinalis 60–61
Carthamus lanatus 62–63
Centaurea cyanus 64–65
Chenopodium album 66–67
Cirsium arvense 68–69
Conium maculatum 70–71

Coreopsis tinctoria 72–73
Cosmos sulphureus 74–75

D
Datura stramonium 76–77
Dianthus deltoides 78–79
Digitalis
    D. grandiflora 80–81
    D. purpurea 82–83
Dysphania ambrosioides 84–85

E
Erigeron
    E. annuus 86–87
    E. canadensis 88–89
    E. sumatrensis 90–91
Eryngium billardierei 92–93
Euphorbia
    E. esula 94–95
    E. lathyris 96–97
    E. prostrata 98–99

G
Galinsoga parviflora 100–101

H
Helianthus tuberosus 102–103
Hemerocallis fulva 104–105

Hesperis matronalis 106–107
Hibiscus trionum 108–109

I

Ipomoea purpurea 110–111
Iris
    I. germanica 112–113
    I. pseudacorus 114–115
    I. reticulata 116–117

L

Lepidium virginicum 118–119
Leucanthemum vulgare 120–121
Linaria dalmatica 122–123
Lupinus polyphyllus 124–125

M

Malva parviflora 126–127
Matricaria discoidea 128–129
Medicago sativa 130–131
Mentha pulegium 132–133
Muscari neglectum 134–135
Myriophyllum aquaticum 136–137

N

Narcissus
    N. poeticus 138–139
    N. pseudonarcissus 140–141
    N. tazetta 142–143
Nymphaea mexicana 144–145

O

Oenothera glazioviana 146–147
Onopordum acanthium 148–149
Ornithogalum umbellatum 150–151
Oxalis
    O. corniculata 152–153
    O. debilis 154–155

P

Papaver
    P. dubium 156–157
    P. rhoeas 158–159
Parthenium hysterophorus 160–161
Populus deltoides 162–163

Q

Quercus robur 164–165

R

Ranunculus
    R. bulbosus 166–167
    R. repens 168–169
Ricinus communis 170–171
Robinia pseudoacacia 172–173
Rubus ulmifolius 174–175
Rudbeckia
    R. hirta 176–177
    R. laciniata 178–179

S

Saponaria officinalis 180–181
Sedum album 182–183
Senecio vulgaris 184–185
Sisymbrium officinale 186–187
Solanum americanum 188–189
Sonchus oleraceus 190–191
Spartium junceum 192–193
Symphyotrichum subulatum 194–195

T

Tagetes
    T. erecta 196–197
    T. minuta 198–199
    T. tenuifolia 200–201
Tanacetum vulgare 202–203
Trifolium dubium 204–205

V

Veronica
    V. hederifolia 206–207
    V. persica 208–209
    V.polita subsp. *lilacina* 210–211
Vinca
    V. major 212–213
    V. minor 214–215

X

Xanthium
    X. spinosum 216–217
    X. strumarium 218–219

Y

Youngia japonica 220–221

Z

Zinnia angustifolia 222–223

If you have any concerns about our products,
you can contact us on
ProductSafety@springernature.com

In case Publisher is established outside the EU,
the EU authorized representative is:
Springer Nature Customer Service Center GmbH
Europaplatz 3, 69115 Heidelberg, Germany

Printed by Libri Plureos GmbH
in Hamburg, Germany